THE PROBLEM OF NATURE

NEW PERSPECTIVES ON THE PAST

General Editor
Constantin Fasolt, University of Chicago

Advisory Editors
Michael Burns, Mount Holyoke College
Pamela Kyle Crossley, Dartmouth College
R. I. Moore, University of Newcastle upon Tyne
Ian Morris, University of Chicago
Patrick Wormald, Christ Church, Oxford
Heide Wunder, Universität-Gesamthochschule, Kassel

IN PRINT

David Arnold The Problem of Nature
Patricia Crone Pre-Industrial Societies
Ernest Gellner Nations and Nationalism
Ernest Gellner Reason and Culture
David Grigg The Transformation of Agriculture in the West
I. G. Simmons Environmental History

IN PREPARATION

Richard Bonney Absolutism
Bernard Crick Representative Institutions
David Gress The Balance of Power
R. M. Hartwell Capitalism
Bruce Trigger Sociocultural Evolution
David Turley Slavery

THE PROBLEM OF NATURE

ENVIRONMENT, CULTURE AND EUROPEAN EXPANSION

David Arnold

BLACKWELL *Publishers*

First published 1996

2 4 6 8 10 9 7 5 3 1

Blackwell Publishers Ltd
108 Cowley Road
Oxford OX4 1JF
UK

Blackwell Publishers Inc.
238 Main Street
Cambridge, Massachusetts 02142
USA

British Library Cataloguing in Publication Data

A CIP catalogue record for this book is available from the British Library.

Library of Congress Cataloging-in-Publication Data

Arnold, David, 1946-
      The problem of nature : environment, culture and European
expansion / David Arnold
          p.   cm. -- (New perspectives on the past)
   Includes bibliographical references and index.
      ISBN 0–631–17732–9. -- ISBN 0–631–19021–X (pbk.)
      1. Environmental sciences -- History.   2. Environmental sciences-
-Philosophy.   3. Natural history.   I. Title.   II. Series.
GE50.A77   1996
304.2'09 -- dc20   96–2054
                                                        CIP

Typeset in Plantin 11 on 13pt by CentraCet Limited, Cambridge
Printed in Great Britain by Hartnolls Limited, Bodmin, Cornwall.

Contents

Preface vii

1 Introduction 1

2 The Place of Nature 9

3 Reappraising Nature 39

4 Environment as Catastrophe 56

5 Crossing Biological Boundaries 74

6 The Ecological Frontier 98

7 The Environmental Revolution 119

8 Inventing Tropicality 141

9 Colonizing Nature 169

Conclusion 188

Further Reading 191

Index 195

Preface

Environmental history has been one of the most rapidly growing areas of historical debate and enquiry in recent decades. Apart from the historian's restless search for new themes and sources, it derives its importance from widely felt concern about the fate of the planet and the consequences of industrial pollution, environmental degradation and climatic change. It is a subject that is clearly very relevant to today's world as well as richly informative about the shaping of times past. But the idea of environmental history contains a basic ambiguity. Should it be the study of what actually happened to the Earth and its creatures over time? Or should it concern itself less with physical reality than with changing perceptions of the natural world and of the human relationship with it? Should it be a branch of natural history, or an essay in historical ideas?

This book is concerned more with the latter than with the former. It is an attempt to show how historians (and other historically minded writers) have invoked nature in its various manifestations – climate, topography, vegetation, wildlife and disease – as a dynamic force in human history. In particular, it tries to show how ideas of geographical and biological determinism have played a major part in historical theory and explanation, and how problematic this invocation of nature can be. The book covers what might increasingly be seen as a whole cycle of environmental history, from the Black Death, the first European voyages of discovery and the rise of capitalism through to the global imperialism of the late nineteenth century and growing concern at the destructive effects of rapid environmental change. Although Europe is in many ways at the heart of this history, the Americas, Asia and Africa form an essential part of environmentalist historiography as it has evolved over the past two centuries. In an important sense, as this book tries to demonstrate, environmental history and its interpretation have long been a central element in the complex material and cultural relationship between Europe and the rest of the world.

The book has grown out of lectures presented in undergraduate history courses at the University of Lancaster and in London at the School of Oriental and African Studies. I am indebted to students on those courses for their often critical response, and also to the stimulus of colleagues with whom I have taught and from whose lectures I have also greatly benefited, particularly John MacKenzie at Lancaster and Gervase Clarence-Smith at SOAS. I have been helped, too, in various ways and at different times by advice and suggestions from Michael Adas, David Anderson, Michael Anderson, Richard Grove, David Hardiman, David Parkin, Peter Robb, Philip Stott, Giles Tillotson, and Elizabeth Whitcombe. I owe a particular debt to Ram Guha, whose own work on the environmental history of India has been a particular source of help and encouragement. Two workshops, one on Indian environmental history held at Bellagio in March 1992, and one on the early history of tropical medicine, held at the Wellcome Institute in London in March 1993, also taught me much and helped me to test some of the ideas presented here. Bob Moore, as the former editor, and Constantin Fasolt, as the present editor, of the New Pespectives on the Past series, and John Davey at Blackwell, also played an invaluable part in getting this book off the ground. As ever, I owe an enormous debt to Juliet Miller for enduring the making of this book over several summers and for helping me to see the wood for the trees.

1
Introduction

Nature, Mr Allnut, is what we are put in this world to rise above.
Katharine Hepburn to Humphrey Bogart in *The African Queen* (1951)

Man is a biological entity before he is a Roman Catholic or a capitalist or anything else . . . The first step to understanding man is to consider him as a biological entity which has existed on this globe, affecting, and in turn affected by, his fellow organisms, for many thousands of years.
Alfred W. Crosby, *The Columbian Exchange* (1972)

Nature might appear not to be a historical problem at all. There have been many historians – past and present – for whom nature has scarcely existed as a significant factor. It has proved perfectly possible to write a history of the French Revolution that made no reference to the climate of France or an account of the rise of Nazi Germany that did not mention German environmentalist ideas. Many historians belong to a tradition, still in many ways the dominant one, in which nature, as either ideology or material reality, does not figure, except perhaps to set the scene for the real drama to follow: the drama of human lives, human action, human-centred events. Indeed, many historians would feel uncomfortable at the intrusion into their narrative of something so abstract or so complex. Nature, they might well argue, belongs to those technically qualified to deal with it: to biologists, climatologists, epidemiologists, and the like. The proper study of history is people.

But of late, whether in the continuing quest for new kinds of history and new types of source materials, or from the stimulus of contemporary environmental concerns, many historians have begun to take a far more positive attitude towards nature's place in the writing of history and the recovery of the past. We live in an age in which nature is conspicuously valued (as pure, wholesome, endangered), even while it is being violated on an unprecedented scale, and it is surely right that historians should address themselves, through their critical understanding of the past, to the concerns that inform and agitate the societies in which they themselves live. But what kind of history should this history of nature or the environment be, and what value might it have in trying to understand and interpret the past?

As in other fields of historical research and debate, there is no single answer to that question: rather, a variety of possibilities presents itself. Nature can be invoked in the form of science, as a source of practical information and authoritative knowledge about the past. The natural sciences can, for instance, tell us about the technical aspects of climate, vegetation and disease, and how these are likely to have had a bearing on the material existence of societies prior to our own. The sciences of nature can also provide us with models of how human societies evolved or responded to the environmental and social crises they faced, and, indeed, to an extent which is perhaps not often realized, the biological sciences have for more than a century had a profound influence on historical thinking. Thus nature as science can provide historians with key ideas about the structure and dynamics of the past as well as yielding technical information for its analysis.

But, rather at variance with this use of science as a source of authority in historical analysis, there has been an increasing disposition among historians to contest the hegemony of science. One can perhaps trace this back to the publication in 1962 of Rachel Carson's *Silent Spring*, with its doom-laden account of the effect of pesticides on animal life and human well-being, and the subsequent growth of environmental awareness in North America and Western Europe. But there has been a more general questioning in recent times of science and technology linked to concerns about industrial and traffic pollution, the dangers of nuclear power and the harmful side-effects or intrusive nature of modern medical and surgical practices. Influenced by this more critical mood, some historians have begun to treat science not as objective fact or authoritative source, but as driven by various political and cultural agendas. Science, it would seem, is too important simply to be left to the technical experts: it needs to be deconstructed and opened up for wider historical scrutiny. In the process we can hope to learn more about the subjectivity of nature, how ideas about the environment have been socially constructed and have served, in different ways at different times, as instruments of authority, identity and defiance.

Environmental history thus deals not just with issues of how the environment has changed (whether as a result of human action or from other causes) and what effects this has in turn had on human societies, but also with ideas about the natural world and how these have developed and informed our very understanding of history and culture. Commonly, the historian is dealing here with several sets of ideas and actions, representing different classes or cultures. One person's wilderness might be another person's Eden. For some people, forests have been home, supplying their every need and comfort; for others, they have been places of darkness and barbarism, fit only to be cut down in the cause of progress, prosperity and good order. The environment has been not just a place, but also an arena in which conflicting ideologies and cultures might become locked in bitter contention.

Although the terms 'ecological history' and 'environmental history' are these days often used synonymously, some distinction might usefully be made between them. The word *oecologie* was coined in the 1860s by a German biologist Ernst Haeckel, though the concept was several decades older and grew out of an earlier tradition of natural history. Haeckel used the term to describe the science of the relations of living organisms to their external world, their habitats, parasites, predators, exposure to certain types of soil, climate, and so forth. Ecology implied a family of living organisms, each dwelling in close proximity to each other, sharing the same physical space, with conflicting appetites or complementary needs. In the main, ecological science has tended to focus on the study of nature as the non-human world and to avoid an anthropocentric approach: it has sought to identify and explain the interrelation of all life forms, and not to privilege the human factor. From a present-day perspective, however, such a stark distinction between people and nature might appear unrealistic: there is not much nature left that is entirely free of some kind of human influence and perhaps there has not been, across most of the globe, for centuries.

Environmental history, by contrast, is more often understood as the story of human engagement with the physical world, with the environment as object, agent, or influence in human history. Here nature figures unashamedly as human habitat, and seasons, soils, vegetation and topography, animal and insect life, are all seen as impinging significantly on human activity, productivity and creativity. By its influence upon land-use and viable modes of subsistence, nature is held to promote or prohibit certain types of social structure, economic organization and even belief systems. To a certain extent this is obvious: life on a fertile, well-watered plain is likely to be very different, culturally and materially, from life in the middle of a desert or on a remote mountain range. The physical features of the environment are taken to be formative influences on the collective identity of any cultural or national group.

In practice, however, the importance given to such environmental factors by historians (as opposed, for instance, to geographers) has varied enormously. For some, the physical environment has been little more than a descriptive aside, a convenient way to begin a book which otherwise deals almost exclusively with political, social or economic developments. But for others, such as those who have written in recent decades about North American Indians or the peasants of medieval Europe, the environment has assumed far greater significance in trying to understand and explain the material life and world-view of those social groups. In such contexts the environment is often seen to be a relatively fixed and stable factor or to give rise to repeated cycles of hunger and plenty. But historians are also concerned with wider processes of change and here, too, the environment has been given a leading role. Some historians have argued that such dynamic environmental factors as climate and disease have had a determining effect on the broad trends of human history, accounting in no small part for the rise and fall of civilizations, the expansion or extinction of entire social and political systems, and the wide disparities between one culture and another. The historical role of biological or environmental determinism, and the way in which the environment has been invoked to explain cultural differences and identities, are two of the principal ways in which nature has been used in the writing of human history, and they are among the main issues discussed in this book.

Although some past expressions of environmental determinism now appear to us crudely simplistic and woefully ill informed, there has been a recurrent fascination among historians (and others outside the historical profession) with the idea of geographical, climatic or biological determinism and the belief that human societies are shaped (and differentiated from one another) by their physical location and environmental circumstances. Until quite recently, the tendency was to see the relationship between people and place in terms of the ascendancy of the latter over the former, the firm hold nature exercised over human lives, or, more cautiously, the limits which the environment set to the possible range of human activities and modes of subsistence. Such a view has often been used to account for the long-term trends and underlying patterns of human history, or to distinguish the past century or two, in which the forces of nature have finally been captured and tamed, from previous ages in which they still held history's reins. This view continues to inform many approaches to history, though perhaps the relationship is now more often seen in an opposing light. An awareness of human subordination to, and dependence upon, nature has a long ancestry, but a sense of human beings as the guardians and destroyers of nature has only recently dawned upon us, and with it an awesome sense of our responsibility for the past destruction and the future survival of other species.

However, even when they turn specifically to the environment itself, historians have differed widely in trying to explain what it is that impels processes of change. Some have found the answer in what one writer (A. W. Crosby) has called 'ecological imperialism'; that is, in the expansionist career of plants, animals and diseases, which moved alongside, or even in advance of, human migration, and wrought fundamental social and environmental change. The Americas following the arrival of Christopher Columbus in 1492 have been taken as a primary example of this, but similar patterns of invasion have been identified for other temperate parts of the globe. This, to be sure, is only one possible line of interpretation. Other historians have seen the driving force behind environmental change and its human consequences to lie elsewhere, in the economic imperatives of capitalism and industrialization, in Europe's relentless quest for land, trade and profit, and in the scientific and technological advances that made the wholesale exploitation of environmental resources both desired and feasible. Others again have stressed cultural values, notably the legacy of attitudes towards nature long established in the West and supposedly originating in the Judaeo-Christian tradition. Historians have contrasted negative attitudes to forests and the wilderness with the quest for a pastoral idyll or an orderly Eden, or have drawn contrasts between environmentally destructive European attitudes and other, more empathetic, cultures in Africa, Asia, America and the Pacific. Perhaps all three lines of interpretation – the biological, economic and cultural – will have a place in any eventual synthesis, but for the present the sheer diversity of these approaches suggests the need for the historian to tread warily and to assess with care the cultural as well as the material aspects of historical change. But such debates also provide us with a specific context within which to discuss the more general problem of nature in history – the expansion of Europe from the fifteenth century onwards.

One of the attractions of environmental history – but also one of its practical problems – lies in the varieties of scale that can be employed. It is possible to write an environmental history which deals with a single country or region. It is perhaps only at the very local level that the historian can satisfactorily identify the many factors – climate, soil, crops, animal life and so forth – that form part of a highly complex environmental story. But at the same time, as talk of 'global warming' and similar planet-wide phenomena remind us, environmental history must also look beyond many of the old geographical boundaries that have long governed our approaches to history. Environmental historians are often aware of patterns of change that affect not just the nation state (around which so much history continues to be constructed) but entire continents. Environmental history has the capacity to be world history in a uniquely dynamic and comprehensive way as diseases, plants and animals move, or are moved, from one continent to another, or as climatic shifts affect regions of the globe far distant from each other. But this too presents problems. Is this a history of increasing global unity or does it in fact reveal a pattern of persistent diversity? On the one hand, historians have drawn attention to a growing unification of the globe by disease, by the transfer of plants, animals, agricultural practices and international trade. But, on the other hand, experiences and expectations of the environment continue to differ widely from one region of the globe to another. Moreover, the perceived importance of the environment as a site of difference (or 'otherness'), between Europe and Asia, or between the tropics and the temperate lands of the 'Neo-Europes', has largely remained, despite the rise of the 'global village'. This, too, is a central theme of the book.

To attempt to cover the whole range of environmental history and to try to address the momentous issues arising from the invocation of nature in history would clearly be impossible in a single, brief work of the present kind. This book is necessarily selective in its focus, but it does aspire to a certain thematic unity. It approaches the problem of nature in relation to European expansion through three interconnected themes: epidemic disease, the moving frontier and the colonization of nature. It concentrates (with a few retrospective glances) on the past five to six hundred years from the early fourteenth century onwards. Many of the most momentous (and certainly many of the most fully documented) instances of environmental change have occurred during the past five centuries, or they repeat (perhaps on a larger scale) episodes from an earlier time. Half a millennium is time enough to gain some sense of environmental change and its historical significance and interpretation. More specifically, the period from the fourteenth to the early twentieth century was an age of growing European economic, political and environmental dominance over the rest of the world, and this was matched by the growth of Western ideas about different environments and the people who inhabited them. Many of the present-day debates about environmental change have clear antecedents in the observations and issues of the past five hundred years or so, and certain historical events and episodes – the Enlightenment, the scientific, agricultural and industrial revolutions, the growth of capitalism and the rise of Western imperialism – fall within this period and mark out important stages in its environmental history.

It will also be suggested here that the period between the Black Death in the 1340s and Columbus's crossing of the Atlantic in 1492 signalled the start of a new ecological age which lasted through to the high imperialism of the late nineteenth and early twentieth centuries. It represents the globalization of certain key environmental factors – disease, for instance – and of certain Western attitudes to the environment. It also represents the transition from what has been seen as Europe's deep environmental crisis in the fourteenth and fifteenth centuries to what appeared to be the technological and ideological mastery of nature in the early twentieth century. The close of the period signifies, too, the beginnings of the first fully articulated doubts about the human impact upon the environment and the first sustained conservationist moves. This great cycle can also be represented, as has been done here, in geographical and cultural terms. The story begins with Europe and European representations of (and responses to) nature, moves through the experience of the Americas (which have, critically, provided a number of central issues and examples for wider historical debate) and then on to the tropics and India, as examples of very different environmental contexts and ideological connections.

2
The Place of Nature

The Environmentalist Paradigm

Although the term 'environment' as we use it today is a relatively recent one, there is nothing new about the idea that the destiny of human beings is intimately bound up with the natural world. But what constitutes 'nature', what effect it has exercised on history, how far history can be written from a biological rather than social or cultural perspective, and what place the environment should occupy in the conceptualization of historical time and space – these are issues that have long been debated and are still very far from being resolved. And even when many of the underlying issues have remained the same, the way in which nature has been invoked in relation to history has changed widely over time as each generation of scholars has reworked the idea of nature to suit its own needs or as historical techniques and perspectives have altered.

The purpose of this and the following chapter is not to discuss the 'idea of nature' as such, nor to examine its historical genesis and cultural significance. That has been dealt with, often very effectively, by other writers. The purpose here is to see writing about the place of nature in history as itself an important subject of historical enquiry. In order to do this, it is necessary at the outset to see 'nature' as not simply something that exists 'out there' – in the lives of plants, the behaviour of animals, or the pattern of winds and ocean currents – but also within our own mental worlds and historical awareness. In order to conceptualize as broadly as possible the ways in which ideas of nature have been invoked in the writing of history, it will be helpful to begin with the 'environmentalist paradigm'.

The environmentalist paradigm provides us with a distinctive model for understanding and explaining the human past. It does not represent nature in the abstract, as an eco-system apart from, or devoid of, human influence and intelligence. On the contrary, it is frankly anthropocentric, seeing in nature a reflection or a cause of the human condition, whether physical, social or moral. It arises from a widely held and historically enduring belief that a significant relationship exists between what is conventionally referred to (even in these gender-conscious times) as 'man' and 'nature', and that this relationship influences the character of individual societies and the course of their histories.

Historically, of course, there has been no consensus as to how this influence might operate. In some instances, it is seen to be so strictly determinist as to allow little room for human free will. The environment, typically in the form of climate and topography, but sometimes also of disease or other 'natural' hazards, dictates the physical and mental characteristics of a society, its modes of subsistence, its cultural life and political institutions. It even determines whether a society is able to scale the heights of civilization or is confined to the depths of savagery and barbarism. Other writers, less given to extremes, have preferred a 'possibilist' position: the physical environment restricts human societies in some ways but not in others, or only does so in the more primitive stages of human development. The more mature and civilized a society becomes, the less it lies in nature's thrall: indeed, the mark of a civilization is precisely its ability to rise above narrow environmental constraints. At other times, and increasingly in recent decades, the basic paradigm has been reversed: mankind has won mastery over nature, it is argued, but has abused and mistreated it, and now must live with the environmental and social consequences of its Promethean act. This kind of environmentalism tends to concentrate on the harm humans have done to the environment (and hence to themselves) through industrial pollution, mechanized farming, the destruction of forests, and the extinction of animal and plant species.

The environmentalist paradigm thus covers a wide range of interpretative options: from history as the harmonious working together of people and nature, at one extreme, to an irreversible ecological crisis precipitated by human greed and folly, at the other. But what is common to all these views is a belief that nature and culture are dynamically linked and that history is in some central way connected with this intimate and continuing relationship.

It is, however, worth reflecting how diversely different ages and societies have understood this environmentalist idea. For instance, barely two hundred years ago it was not uncommon in Western society to stress the harmful influences of the environment, one in which disease-bearing 'miasmas' or 'foul exhalations' were believed to issue from almost every possible source – from marshes and swamps, from river banks, forests and dense undergrowth, as well as from human habitations. Although the term as such was not then used, environmentalism at that time signified an awareness of the hazards of nature and the need, for the sake of one's health, to avoid them as much as possible or to destroy the source of the miasmas by draining marshes and clearing jungles. Today, Western attitudes have been substantially reversed. The 'natural' environment is seen to be intrinsically healthy. It is we who make it unhealthy by our abuse of it or endanger our health by interfering with nature.

But, despite these major and often rapid interpretative shifts, it can none the less be argued that, while the form in which the environmentalist paradigm is expressed has changed significantly over the centuries, there remain a number of common and recurring elements. One is the close physical and cultural affinity seen to exist between nature and humankind. As the example just given indicates, this connection has often been understood in terms of health and disease, the well-being or sickness of the one (nature) being reflected in the vitality or morbidity of the other (humankind). There was an influential medical strand in environmentalist writing from the ancient Greeks through to the eighteenth century (and well beyond). The triad of climate, health and medicine has arguably been one of the most powerful and pervasive idioms in which the environmentalist idea has historically been expressed. And it is a mirroring of humankind and nature that finds close parallels in other cultures, in the Hindu Ayurvedic system of medicine, for instance, or Chinese geomancy.

A second feature is the way in which the environmentalist paradigm has been used to establish 'otherness', to make contrasts between different societies as well as to explain the cultural and historical idiosyncrasies of any one society. Hence ethnography occupies a place alongside medicine among the diagnostic tools of environmental determinism. The comparison of cultures across geographical space, as well as through historical time, has been one of the commonest uses of the environmentalist paradigm. Environmentalism has often risen to prominence as an explanatory mechanism in times of rapidly widening geographical horizons or intensified inter-ethnic contact resulting from trade, migration, conquest and colonization. Environmentalist ideas have served at various times to explain perceived differences between supposedly civilized and savage peoples, between the temperate and tropical zones, or between east and west (as between Europe and Asia, or Eurasia and America).

The environmentalist paradigm has thus served to articulate not only the kind of historically and culturally constructed relationship between humankind and nature that has been the subject of so much environmental history but also a relationship of actual or incipient power and authority between one set of human beings and another. If one reason why environmental ideas have been repeatedly invoked has been the attempt to find a coherent pattern in the course of human history – to give it an underlying structure and to see it as more than a random sequence of events – another has been to find an appropriate basis for comparison between the cultural forms and historical evolution of one society and another. What is it, historically, that societies divided in time and space share? What is it, environmentally speaking, that keeps them apart and drives them along such different paths?

The environmentalist paradigm is, of course, only one of a number of explanatory devices that have been used to explain the course of history. Karl Marx located the dynamics of human history not in the dialectic of humankind and nature, but in dialectical materialism, in successive modes of production, such as feudalism and capitalism, and in the struggle of class against class. It should be noted, though, that while Marx was intent on uncovering the universal laws of history and society and not in exploring local particularisms, the so-called 'Asiatic mode of production' (with its stress upon the physical peculiarities of Asian societies) left the door ajar for at least one form of environmental determinism to enter Marxist thought. In general, though, nature, where it figured at all in the writings of Marx and Engels, did so in a strictly subordinate and instrumentalist role. The 'subjection of nature's forces to man', noted in the *Communist Manifesto* of 1848, was but one of several expressions of the triumphant power of the bourgeoisie, a measure of its mastery over the means of production as a whole. Capitalist or communist, humankind in the Marxist view was 'the sovereign of nature', not its subject.

Other writers, less so now than in the nineteenth and early twentieth centuries, sought their explanatory mechanism in race, in the supposed mental, physical and moral superiority of one race over others or in the racial origins of certain social and political institutions. At various times and in many cultures, divine will (or Providence) has served to explain an otherwise inexplicable course of events and to give meaning and purpose to life on Earth. In the late twentieth century, gender has been used to explain many of the critical social, cultural and political developments of history. These sets of determining forces – environment, class, race, divine will, gender – have all, in their various ways, been used in attempts to mediate between nature and culture, between the physical (whether represented by climate, soil or human biology) and the cultural aspects of human existence. Is humankind an integral part of nature, an essentially biological being, or have we been set apart (or grown apart) from nature – whether by God's will or by our own laws and customs?

At times these contending paradigms have fought bitterly as embattled adversaries, alternative ways of seeing, understanding and representing the world. The decline, or unpopularity, of racial and providential interpretations of history over the past fifty years has perhaps been one reason why environmentalist explanations have come increasingly to the fore. But because of its elasticity as an explanatory device, or possibly because a troubled sense of our relationship with nature lies so deep in our culture and unconscious, environmentalist ideas have also freely combined with other explanatory paradigms. One can see this in nineteenth-century arguments about the interdependence of race and environment (a linkage discussed later in this chapter and elsewhere in the book). There is also the more recent example of eco-feminism, which has sought to identify environmental control and degradation with women's subordination and exploitation in a combined critique of patriarchy, capitalism and science.

Airs, Waters, Places

Ideas about the influence of the environment on human culture and physique have a long history in Western thought, a history explored in detail by Clarence J. Glacken in his pioneering study *Traces on the Rhodian Shore*.[1] There is no point in covering the same ground here, but in order to establish some of the fundamental tenets of the environmentalist idea, it is helpful to begin where Glacken does, in classical Greece, and with the text known as *Airs, Waters, Places*. Attributed to the physician Hippocrates of Kos in the fifth century BC, this brief treatise falls into two parts: one medical, the other ethnographical. The two parts may originally have been separate works by different authors, but, even if

[1] Clarence J. Glacken, *Traces on the Rhodian Shore: Nature and Culture in Western Thought from Ancient Times to the End of the Eighteenth Century*, Berkeley, 1967.

their union was fortuitous, their combined effect has been of lasting significance for environmentalist ideas.

The first part establishes the importance of medicine and physiology in relation to the environment. Its avowed purpose is to help a doctor understand the causes of the diseases he is likely to encounter in moving to a new locality – a district, for instance, that is exposed to northerly winds, or that draws its water from stagnant ponds and marshes rather than from free-flowing springs, a place where the seasons, soils and vegetation may be very different from those with which he is familiar. By knowing such things, the physician will not only maintain good health himself but be able to give his patients appropriate guidance. It is assumed that all human beings are basically much alike: what makes them different are the environmental forces, the airs, waters and places, and hence the diseases, to which they are exposed. In Greek thought, the human body was often perceived as a microcosm of nature, and thus what agitated nature, like a cold north wind or the change from season to season, was bound to have a corresponding effect upon human physiology and on the four bodily humours, or fluids, which determined susceptibility to disease.

The second part of *Airs, Waters, Places* begins abruptly: 'I now want to show how different in all respects are Asia and Europe, and why races are dissimilar, showing individual physical characteristics.' The contrast between Europe and Asia (in the restricted sense used by the Greeks), and the superiority of temperate climates, were themes that resurfaced in later environmentalist writing from the seventeenth century onwards, often in much the same words as Hippocrates used. He observes:

Asia differs very much from Europe in the nature of everything that grows there, vegetable or human. Everything grows much bigger and finer in Asia, and the nature of the land is tamer, while the character of the inhabitants is milder and less passionate. The reason for this is the equable blending of the climate, for it lies in the midst of the sunrise facing the dawn. It is thus removed from extremes of heat and cold. Luxuriance and ease of cultivation are to be found most often when there are no violent extremes, but when a temperate climate prevails.[2]

This part of the text is not so consistently determinist as the first. Hippocrates gives scope for cultural factors in the making of social as well as physical characteristics, such as among the Macrocephali, whose custom of moulding and binding the heads of their infant children is said to have given them their distinctive shape. But the influence of climate and other environmental factors is seldom absent from the discussion. 'Climates differ,' it is stated, 'and cause differences in character; the greater the variations in climate, so much the greater will be differences in character.' As a rule, 'the constitutions and the habits of a people follow the nature of the land where they live.' Thus lands that boast rich, well-watered and easily cultivated soils, and that are not subject to great variations of climate, produce lazy, cowardly people, incapable of hard, physical work and little disposed to mental exertion either. By contrast, where the land is bare, waterless and rough, swept by winter winds and burnt by summer sun, the inhabitants are hard and spare, keen in intellect, skilled in crafts, brave and proficient in the arts of war.[3]

These environmental differences and their human consequences are again used to make a contrast between the peoples of Asia and Europe:

> The small variations of climate to which the Asiatics are subject, extremes both of heat and cold being avoided, account for their mental flabbiness and cowardice as well. They are less warlike than Europeans and tamer of spirit, for they are not subject to those physical changes and the mental stimulation which sharpen tempers and induce recklessness and hot-headedness. Instead they live under unvarying conditions. Where there are always changes, men's minds are roused so that they cannot stagnate.[4]

[2]   G. E. R. Lloyd (ed.), *Hippocratic Writings*, Harmondsworth, 1983, p. 159.
[3]   Ibid., pp. 161, 168–9.
[4]   Ibid., p. 160.

While environmental factors are seen largely to account for 'the feebleness of the Asiatic race', a further cause is said to be that much of Asia is under monarchical rule. Under a monarchy, says the text, subjects are compelled to fight though this serves only their master's glory; whereas when men govern themselves, under a democracy, as in parts of Greece, they reap the rewards of their own valour and so have more incentive to fight bravely.[5]

The significance of *Airs, Waters, Places* does not lie in its contribution to Greek thought alone, though Aristotle and the historian Herodotus (among others) were either influenced by it or shared largely similar views. As Glacken puts it, the treatise represents 'the first formulation of the environmental idea' that human minds, bodies, even whole societies, were shaped by their geographical location, their climate and topography. In particular, the idea that equable climates and fertile soils produce lazy people, and bleak, barren lands brave men persisted in European thought well into the nineteenth century, and has perhaps not entirely disappeared. But *Airs, Waters, Places* is significant for our discussion in at least two other respects. First, as an early exercise in comparative ethnography, it purports to explain why the peoples of the world, as then known to the Greeks, were so different from one another: the Egyptians and Libyans at one end, the Scythians of the Russian steppe at the other, with the Greeks themselves ideally located somewhere in the middle. Characteristically, the text identifies Europe as the norm and Africa and Asia as the aberrant extremes, again a recurrent motif in European environmentalist thought down the ages. Even at this early stage in its evolution, the environmentalist paradigm was used not just to explain why a particular people, such as the Greeks, were as they were, but why other sets of people differed so radically from them in their appearance, manners and customs.

Secondly, it is important that the author of *Airs, Waters, Places* was believed to be Hippocrates, a physician, and that the first half of the treatise deals specifically with the relationship between

[5] Ibid.
[6] Glacken, *Traces on the Rhodian Shore*, p. 87.

human health and the surrounding forces of nature. In the Greek system of medical thought, the latter were believed to disrupt the balance of bodily humours necessary to maintain good health: they brought, according to their fashion, an excess of phlegm or a surfeit of bile. Illnesses and conditions of all kinds, from swelling of the spleen and eye diseases to convulsions, miscarriages, dysentery and pleurisy, were thus taken to be the physical manifestations within the body of the malign influence of chill winds, brackish water or a sudden seasonal change in temperature. As already noted, this represents a widely held belief in the linkage or analogy between human health and the health or stability of the surrounding environment. Our own experience or perception of sickness and health is one of the principal standpoints from which we evaluate our environment or the environment in which other people live. This health/environment connection is certainly to be found in many different ages and cultures, but its association with Hippocrates, the archetypal physician of Western thought, was extremely influential in shaping and maintaining the environmentalist idea in the West. The physician is respected as both an experienced observer of nature and as a mediator seeking to restore harmony between the patient and the physical environment.

'The Most Powerful of All Empires'

But we cannot tarry too long among the Greeks. If we leap forward two millennia from ancient Greece to the eighteenth century of the Christian era, we find ideas of environmental determinism not unlike those expressed in *Airs, Waters, Places* and, indeed, drawn directly from that Hippocratic text. Environmentalist ideas enjoyed a particular vogue in the eighteenth century, and more generally from the middle of the seventeenth century through to the late nineteenth. This was evident across a wide cultural and social spectrum – in medicine and science, in philosophy and aesthetics, in painting, poetry and even in landscape gardening. It was as if eyes, long half-closed, were suddenly opened and Western society saw nature, for the first time, as complex and glorious. 'In no preceding age,' Glacken remarks, 'had thinkers discussed questions of culture and environment with such thoroughness and penetration as did those of the eighteenth century.'[7] The idea of nature, it has further been said, was '*the* governing idea of the age of the Enlightenment'.[8]

It is more difficult, however, to pinpoint why this was so. The new 'feeling for nature' arose from a number of causes. Advances in physics, astronomy and botany since the sixteenth century had given a greater understanding of the forms and effects of the natural world and prompted an unprecedented desire and ability to use and control the forces of nature. The new-found security and affluence of rulers and aristocrats, with their country seats and patronage of the arts and sciences, contributed to the growing interest in nature, as did the rise of rural capitalism and a new sense of pride and proprietorship over an 'improved' and well-managed landscape. Conversely, urbanization and the beginnings of industrialization stirred a Romantic reaction, feeding an appetite for the attractions of pastoral idylls or for wild mountain scenes and tempestuous seas. Much credit is given to Jean-Jacques Rousseau for giving expression to a new sensitivity to nature, a nature that spoke to the imagination as much as to the senses and that served to contrast the simple life of the 'noble savage' with the artificiality and constraints of Europe's courts and cities. What Rousseau and other philosophers, scientists, painters and poets did was to make nature more than simply a subject of intellectual enquiry. It also became one of the principal metaphors of the age, the prism through which all manner of ideas and ideals were brilliantly refracted.

As in the age of Hippocrates, Europe's expanding geographical horizons were an important stimulus to environmentalist thought. A new phase of European contact, exploration and discovery in

---

[7] Ibid., p. 501.

[8] J. Ehrard, *L'Idée de nature en France dans la première moitié du XVIIIe siècle*, 1963, quoted in D. G. Charlton, *New Images of the Natural in France: a Study in European Cultural History, 1750–1800*, Cambridge, 1984, p. 7.

Asia, Africa, America and the Pacific – epitomized by Captain James Cook's three Pacific voyages between 1768 and his death in 1779 – encouraged fresh observation and activity in fields as diverse as painting and horticulture, philosophy and ethnography. The scope for the discussion of the impact of environmental forces on society, and, conversely, the environmental changes resulting from human action, had suddenly and dramatically widened. In medicine, a key science and professional practice in an age when dedicated scientists were rare, the revival of Hippocratic thought brought renewed emphasis upon the interdependence of climate, topography and health. The attribution of many diseases to poisonous miasmas, arising from tainted soil and swamps, like the new-found enthusiasm for spas, sea breezes and coastal resorts, was a further sign of an intense preoccupation with both the malign and beneficial effects of the environment. Looking across the great expanse of the arts and sciences, the period in Britain from the founding of the Royal Society of London in 1662 to the Great Exhibition of 1851 has surely a claim to being an environmentally conscious age to a degree that our own 'age of ecology' (to use Donald Worster's phrase)[9] has barely begun to equal.

There are many poets, painters, scientists and philosophers who could be used to illustrate the ways in which the environmentalist paradigm was expressed in the eighteenth century. But there is perhaps one who stands out as both capturing the spirit of his environmentalist age and having a profound impact on the generations that followed: Charles Secondat, Baron de Montesquieu, whose *L'esprit des lois* (*The Spirit of the Laws*) was published in 1748. In this highly influential work, Montesquieu sought to find a common pattern in the diverse laws that regulated societies that had flourished in the past or existed in his own age. Although his methods were inconsistent, he did this partly by relating law and custom to the physical environment. Through a series of wide-ranging historical and geographical case studies, that com-

[9] Title of the 'Epilogue' in Donald Worster, *Nature's Economy: a History of Ecological Ideas*, Cambridge, 1985, pp. 339–48.

bined classical antiquity with the burgeoning travel literature and ethnography of his own day, Montesquieu tried to show how cultural characteristics and social formations were related to climate and geographical location. Everything from human physiology to religion and morality was, he believed, conditioned by topography and climate. For Montesquieu, as for many of his contemporaries, 'the empire of the climate' was 'the first, most powerful, of all empires'.

Montesquieu repeated the Hippocratic formula that fertile soils produced weak, cowardly men, while barren soils bred brave sons. Warm climates relaxed the body; cold ones were bracing. It followed that 'people are, therefore, more vigorous in cold climates.' Nowhere was this principle of environmental determinism more clearly exemplified than in the case of Asia. As with Hippocrates two thousand years earlier, Asia was invoked by Montesquieu to point an instructive contrast with Europe, but by the eighteenth century its value as antithesis, as Europe's 'other', had been enhanced by the long struggle against Islam and the Ottoman Turks in the Balkans and the eastern Mediterranean. Montesquieu differed, however, from Hippocrates in arguing that Asia had no true temperate zone, and believed that the extremes of climate and topography to be found there – from the severe, cold and barren wastes of its northern regions (Siberia, Mongolia and Manchuria) to the steamy fecundity of its southern zone (Persia, India, China) – gave rise to corresponding extremes in government. Nature dictated that Asia should produce autocratic systems of government: it was 'that region of the world where despotism is so to speak naturally domiciled'. By contrast, it was the milder, more equable climate of Europe, and the more diverse and fragmented nature of its terrain, that resulted in its more moderate laws and more balanced system of government.

Asia [Montesquieu averred] has always been the home of great empires; they have never subsisted in Europe. For the Asia of which we know has vaster plains than Europe; it is broken up into greater masses by the surrounding seas; and as it is further south, its springs run more easily dry, its mountains are not so covered with snow, and its rivers are lower and form lesser barriers. Power therefore must always be despotic in Asia, for if servitude were not extreme, the continent would suffer a division which the geography of the region forbids.[10]

In Europe, by contrast, 'the natural dimensions of geography' favoured the creation of states 'of a modest size', their survival dependent upon the rule of law. There existed a 'spirit of liberty' which rendered each part of the continent 'resistant to subjugation or submission by a foreign power'. But in the countries of Asia there reigned 'a servile spirit, which they have never been able to shake off'. In the entire history of Asia, it was 'impossible to find a single passage which discovers a spirit of freedom; we shall never see anything there but the excess of slavery.'[11]

However, for all his critics' scorn, Montesquieu was not uniformly deterministic and did not see climate as the only possible factor involved in fashioning human laws and institutions. Indeed, he saw it as the task of good law-makers and good governments to rise above the constraints imposed by climate and legislate for their subjects' needs, just as barren soils might, through diligent toil, be made to yield 'what the earth refuses to bestow spontaneously'. 'The more the physical causes incline mankind to inaction, the more the moral causes should estrange them from it,' he wrote. Thus, despite many statements seemingly to the contrary, Montesquieu showed that nature might be permissive and not merely prohibitive, that morality as well as climate might shape human society and endeavour. Control over water and the modifications of nature this entailed seemed, from the examples of societies as far apart as Holland, Egypt and China, to illustrate the potential of humanly modified landscapes. 'Mankind by their industry, and by the influence of good laws, have rendered the earth more proper for their abode,' Montesquieu declared. 'We see rivers flow where there have been lakes and marshes.' This

---

[10]   Cited in Perry Anderson, *Lineages of the Absolutist State*, London, 1979, p. 465.
[11]   Montesquieu, *The Spirit of the Laws*, first published 1748, New York, 1975, p. 269.

was a 'benefit which nature has not bestowed', though it was still 'a benefit maintained and supplied by nature'.[12]

Montesquieu was partly able to advance his environmentalist arguments because so little was then known about climate and its effects, or about the nature and evolution of the societies, past and present, that he presumed to write about with such assurance. 'Ignorance,' Glacken remarks of the climatic theories of Montesquieu and his contemporaries, 'permitted wider generalization.'[13] Nor was Montesquieu a particularly original thinker. Many of his determinist ideas had, for example, been anticipated by Jean Bodin a century and a half earlier. His importance lay in his ability to absorb and synthesize the kinds of environmentalist ideas that were circulating in Europe, to present them in an attractive and relatively coherent form, and to produce some forceful moral conclusions. Not every writer agreed with Montesquieu's more extreme views, and many of his contemporaries preferred to see climate as but one factor affecting human character, or of more importance in relation to savage than civilized societies, but on the whole *The Spirit of the Laws* reflected and strengthened the environmentalist obsession of the age.

It has been said that Montesquieu was greatly influenced in his views of a 'timeless' and 'unchanging' Asia, with its supposed absence of private property in land and its despotic rulers, by the writings of François Bernier, a Frenchman who travelled extensively in India, Persia and the Ottoman domains in the seventeenth century and was at one time physician to the Mughal emperor Aurangzeb. But similar views are to be found in the writings of many other writers of the period, including Sir John Chardin whose *Travels into Persia and the East Indies* were published in 1686, or Tavernier, another French traveller in Mughal India. The growing accessibility of the printed word made such accounts readily available to Europe's armchair philosophers. Just as European merchants and financiers were struggling to bring all the world's lucrative trade within their grasp, so European intel-

---

[12]   Ibid., pp. 273–4.
[13]   Glacken, *Traces on the Rhodian Shore*, p. 620.

lectuals were struggling to bring their fragmentary knowledge of the outside world into a coherent system of order and control. Montesquieu illustrates the apparent utility of the environmentalist approach in trying to incorporate and systematize knowledge about such diverse and relatively unfamiliar societies as India, China and North America. The belief that such societies could only be understood within the context of their physical environments, and that these shaped their moral and material characteristics to a degree unmatched in Europe itself, was a recurrent theme in Enlightenment thought, pursued alike by philosophers and practising administrators. The most celebrated attempt was the *Encyclopédie* in France between 1751 and 1765, but no less significant were the innumerable manuals, gazetteers, statistical surveys and census reports, inspired by the same encyclopedist ideal, produced in French-occupied Egypt or in India under the British. There, too, accounts of climate, topography, vegetation and soil typically framed (and set the parameters for) discussions of local history, culture and ethnography.

The kinds of environmentalist ideas developed by Montesquieu in relation to the distinction between Europe and Asia became 'a central legacy for political economy and philosophy thereafter',[14] and found a place in such seminal works as Adam Smith's *Wealth of Nations* and Hegel's *Philosophy of History*. The claim that for climatic and topographical reasons Asia was 'timeless' and 'unchanging', that it lacked private property in land, and was invariably subject to autocratic rule, also found its way into the writings of Marx and Engels and their much-debated references to 'oriental despotism' and an 'Asiatic mode of production'.

In the century or so following the publication of *The Spirit of the Laws* there were countless other writers – historians, geographers, medical topographers, colonial ethnographers – who eagerly pursued Montesquieu's environmental determinism and, through their own work, sought to provide evidence of its universal validity. Most tended to argue, even more than their mentor, that while other societies had been moulded by their environments and were

[14] Anderson, *Lineages of the Absolutist State*, pp. 465–6.

still governed by them, Europe by virtue of its intellect and industry had uniquely broken free of the shackles of climatic constraint.

One example that was widely read and extensively cited, partly because it was the work of a historian, was H. T. Buckle's *History of Civilization in England*. It was published in three volumes in 1857–61, a century after *The Spirit of the Laws* and in an age when Europe's mastery over nature was being ever more confidently proclaimed. Despite the book's title and seemingly narrow focus, Buckle felt unable to embark on his allotted task without first expounding on the effects of certain 'physical laws' – such as climate, soil and natural disasters – on the development of civilization. Typically, he illustrated his thesis with the age-old contrast between Europe and Asia. Civilization in Asia, he maintained, had a number of natural advantages, particularly the abundance of fertile soil contained in its vast river basins and deltas. Europe was less favourably endowed, but there, unlike in Asia, the 'determining cause' of civilization was 'not so much these physical peculiarities, as the skill and energy of man. Formerly the richest countries were those in which nature was most bountiful; now the richest countries are those in which man is most active.' Europe had learnt, as Asia apparently had not, how to compensate for nature's 'deficiencies'. 'From these facts,' Buckle argued, 'it may be fairly inferred, that the advance of European civilization is characterized by a diminishing influence of physical laws, and an increasing influence of mental laws.'[15] Thus was Europe's growing sense of superiority and strength relative to Asia and the rest of the world measured in terms of its unique ability to surmount and subordinate the forces of nature.

'Race is Everything'

By the time that Buckle was writing in the 1850s, the popularity of such crude ideas of climatic and geographical determinism as

---

[15]   Henry Thomas Buckle, *History of Civilization in England*, London, 1883 edn, vol. I, p. 156.

Montesquieu had espoused was beginning to wane and to be pushed aside or subsumed within a new paradigm – race. The nineteenth- and early twentieth-century preoccupation with race as a way of explaining the dynamics of history and culture arose from several causes. First, the issue of slavery and its abolition sparked intense debates on both sides of the Atlantic over whether Africans belonged to a distinctive – and, it was assumed, inferior – human sub-species. Secondly, Europe's growing military and economic ascendancy was taken as a sign that Europeans were racially superior as well, especially when their arrival in many parts of the world was followed by the precipitous decline and even extinction of indigenous peoples. Thirdly, the eighteenth and early nineteenth centuries saw a rapid growth in the biological sciences, which in turn fostered interest in the differences between races and between human beings and the rest of the natural world.

By the middle decades of the nineteenth century, all these factors were combining to produce a new and aggressive idea of race. Increasingly, race was viewed in competitive as well as evolutionary terms, a dual development aided by the publication of Charles Darwin's *On the Origin of Species* in 1859. With his emphasis on the perpetual contest between species and the 'survival of the fittest', Darwin appeared to reject the idea of nature as fixed, harmonious and God-given. Darwin's evolutionary ideas were readily applied to humans and used to support the view that different races represented different stages in the evolutionary process and that different environmental conditions had been a significant factor in this diversification.

By the mid-nineteenth century, science and history were also beginning to influence and imitate each other. In rejecting a relatively recent date for the Creation and by looking to the testimony of rocks and fossils, scientists had begun to discover time, albeit time on a far more extended scale than historians employed. Charles Lyell's *Principles of Geology*, the first volume of which appeared in 1830, and the evolutionary biology of Darwin's *Origin of Species* challenged the conventional view that the Earth was only a few thousand years old and that its creatures had been fixed and unchanging from the dawn of time. Instead, it was argued that nature was not immutable, but had changed continuously over the vast aeons of the Earth's history, as whole continents and oceans rose and fell, as climates oscillated between desert heat and arctic cold, and as trilobites and dinosaurs died out and other species scrambled to occupy their ecological niches. Biology and geology came to resemble history in their concern for processes of change and the sequential effects of time: indeed, 'what Darwin did was to take into science a way of reasoning invented by historians.'[16] Analogy was no longer made between humankind and machines, as it had been in the age of Newton and Descartes, but between 'the processes of the natural world as studied by natural scientists and the vicissitudes of human affairs as studied by historians'.[17] It was a two-way traffic. Historians, in pursuit of their own quest for authority and meaning, sought to give scientific authenticity to their work by importing into the history of races and nations ideas of evolution and the 'survival of the fittest' drawn from Darwinian science. Civilizations were not immutable species: they evolved or decayed in response to certain environmental conditions; they battled, like dinosaurs, for supremacy and survival.

Race became an obsession, as climate had been a century earlier. It was similarly used to explain all manner of historical and cultural circumstances and to provide a universal framework for the incorporation of the ever-expanding knowledge of the world outside Europe. In 1874 Benjamin Disraeli remarked in his novel *Tancred*, 'All is race: there is no other truth.' Three years later, Robert Knox, one of the most ardent and influential of the Victorian race theorists, was even more epigrammatic, declaring simply: 'Race is everything.'[18]

And yet, while race was often regarded as a self-sufficient and self-evident dynamic, used to explain and justify the superiority

[16] David Knight, *Ordering the World: a History of Classifying Man*, London, 1981, p. 16.
[17] R. G. Collingwood, *The Idea of Nature*, Oxford, 1945, p. 9.
[18] Quoted in Michael D. Biddiss (ed.), *Images of Race*, Leicester, 1979, p. 12.

of Europeans on a global scale, geographical and climatic determinism was also used to bolster racial arguments. One prominent
statement of the environmental underpinning of human evolution
was made by the naturalist Alfred Russel Wallace in 1864. He
argued that although the Earth may once have been inhabited by
a homogeneous human race, as it spread out across the globe it
came under the influence of different environments. Nature's
diversity left its imprint in racial differences. Harsh soils and
inclement seasons were a stimulus for the evolution of a hardier,
more provident and social race (the Europeans) than those who
lived in warmer climates with year-round abundance. Wallace
made a sweeping claim for the superiority of races that had
evolved in the temperate zone when he asked:

> Is it not the fact that in all ages, and in every quarter of the globe,
> the inhabitants of temperate have been superior to those of tropical
> countries? All the great invasions and displacements of races have
> been from North to South, rather than the reverse; and we have no
> record of there ever having existed, any more than there exists
> today, a solitary instance of an indigenous inter-tropical
> civilization.[19]

Wallace, with Darwin the co-author of the theory of 'natural
selection', argued that the 'great law' of the 'preservation of
favoured races in the struggle for life' would inevitably lead to the
extinction of 'all those low and mentally undeveloped populations
with which Europeans come in contact'. Alluding to a pattern of
rapid demographic decline among indigenous peoples following
the arrival of Europeans, Wallace alleged that the North American
Indians, the Indians of Brazil, the Australian and Tasmanian
Aborigines and the Maori of New Zealand 'die out, not from any
one special cause, but from the inevitable effects of an unequal
mental and physical struggle. The intellectual and moral, as well
as the physical qualities of the European are superior . . . .'

[19]   A. R. Wallace, 'The origin of human races and the antiquity of man deduced
from the theory of "natural selection"', *Journal of the Anthropological Society*,
1864, in Biddiss (ed.), *Images of Race*, p. 47.

Significantly, too, in the light of both Victorian evolutionary thought and more recent historical scholarship, Wallace saw this process of depopulation and the displacement of southern peoples by a superior northern race as having an exact parallel among animals and plants. A superior physique, intellect and culture enabled the European to conquer the 'savage', and

> to increase at his expense, just as the more favourable increase at the expense of the less favourable varieties in the animal and vegetable kingdoms, just as the weeds of Europe overrun North America and Australia, extinguishing native products by the inherent vigour of their organization, and by their greater capacity for existence and multiplication.[20]

Thus was racial superiority among humans seen to have a direct counterpart in nature and imperialism vindicated by reference to a parallel process of conquest and colonization among plants and animals.

Wallace was no historian, but his references to civilization – or the want of it in the tropics (a theme we will return to in chapter 8) – show how readily Victorian writers equated the empire of nature with human imperialism. It was not surprising that many historians of the period imbibed such scientific ideas of biological determinism and incorporated them into their own work, or through their historical writing further publicized and diffused the belief that humankind was involved, like plants and animals, in a similar struggle for the 'survival of the fittest'.

Civilization and Climate

By the close of the nineteenth century, environmentalist ideas were being refashioned to meet the ideological imperatives of a new imperial age, and not only by naturalists, anthropologists and historians. Geographers, too, were actively involved, spurred on by an emerging sense of professionalism and impelled by involve-

[20]   Ibid.

ment in empire as a geographical project. Perhaps it is true that geographers have always had a strong belief that human beings are 'creatures of their environment',[21] but the heady combination of racial Darwinism with an ascendant Western imperialism thrust environmentalist ideas into a position of exceptional influence and prominence between the 1890s and late 1920s.

Significantly, in view of its own rise to imperial power, the United States produced some of the most emphatic statements of geographical determinism during this period. In 1911, for instance, Ellen Churchill Semple published a widely cited study of the influence of the 'geographic environment'. Her book opened with a resounding declaration of the supremacy of nature.

> Man is a product of the earth's surface . . . the earth has mothered him, fed him, set him tasks, directed his thoughts, confronted him with difficulties that have strengthened his body and sharpened his wits, given him his problems of navigation or irrigation, and at the same time whispered hints for their solution. . . . Man can no more be scientifically studied apart from the ground which he tills, or the lands over which he travels, or the seas over which he trades, than polar bear or desert cactus can be understood apart from its habitat.[22]

Invoking Montesquieu, Buckle, but above all Darwin, Semple sought to illustrate this determinist hypothesis with examples from across the globe and from different stages of human history.[23] Thus she compared the history of Britain and Japan, on the grounds that both were islands and so exposed to similar environmental – and hence historical – influences. She attempted to define the entire history of a country like Spain solely in terms of its geographical characteristics. What could not easily be explained by geography was accounted for by race, though usually

---

[21] Margaret T. Hodgen, *Early Anthropology in the Sixteenth and Seventeenth Centuries*, Philadelphia, 1964, p. 288.

[22] Ellen Churchill Semple, *Influences of Geographic Environment on the Basis of Ratzel's System of Anthropo-Geography*, London, 1911, pp. 1–2.

[23] Ibid., pp. 12–13.

the one was seen to reinforce the other. A strong strand of Darwinism ran through her account, as in describing the struggle for land between the immigrant and indigenous peoples: 'The struggle for land means a struggle also for the best land, which therefore falls to the share of the strongest peoples. Weaklings must content themselves with poor soils, inaccessible regions of mountain, swamp or desert. There they deteriorate, or at best strike a slower pace of increase or progress.'[24]

Even in her own lifetime Semple's work was widely criticized for drawing extreme conclusions from random evidence and for gross insensitivity to historical change. How could the same island geography and temperate climate account for every twist and turn of British history from the Anglo-Saxons to Magna Carta, or from Cromwell to the defeat of Napoleon? None the less, Semple, like Montesquieu, 150 years earlier, had given expression to ideas and attitudes that were widely felt. There were other historically minded geographers who drew similar conclusions. The most influential of these was another American, Ellsworth Huntington.

In his early works, written in the first twenty years of the twentieth century, Huntington attempted to show how climate profoundly affected human history. Like so many environmentalists before him, he first looked east, seeking in climate and climatic change an explanation for the differences between Western dynamism and Eastern stagnation. In 1907 he postulated that the rise and, more especially, the fall of civilizations was closely linked to climatic change, and to associated factors like endemic hunger and disease. Writing at the close of a period which had seen many Asian countries suffering from drought and famine, Huntington stressed the role of deficient rainfall and hunger in producing a 'state of irritability' in society. Over a long period of time the pressure of hunger and the resultant unrest and insurrection must, he argued, generate a permanent instability in Asian societies and so inhibit their capacity for stable government and civilized life. Huntington did not claim that Western societies were entirely immune from the effects of climatic variability:

[24] Ibid., p. 113.

indeed, he recalled that 'during the years when Chinese Turkestan had its migrations and rebellions, Persia its famines, and Turkey its revolts and massacres', the United States, too, had experienced financial panic and political unrest. If even the stability of the United States could be affected by temporary climatic change and consequent poor harvests, how much greater, he reasoned, must the effect of drought and famine be upon generation after generation of Asiatics?

In earlier, pre-Lyellian times, environmentalist writers had assumed that climate, apart from the cycle of the seasons, was constant. Indeed, one of the criticisms made of Montesquieu had been that it was difficult to explain the decline of ancient Greece in environmental terms when the climate was the same as when Greek civilization had been at its height. But Huntington gave a new twist to the determinist tale (and one taken up by many climatologists and physical geographers since) by arguing that climate *had* significantly changed in historical times with the result that many once stable societies had been unable to adapt and survive. 'Everywhere in arid regions we find evidence,' he wrote, 'that dessication has caused famines, depopulation, raids, wars, migrations, and the decay of civilization.' Where once mighty empires had ruled – as in Central Asia – there now only roamed bands of nomadic pastoralists. Even ancient Rome, for all its achievements, had not been immune to climatic change and had fallen prey to the consequent spread of food shortages and malaria. Long-term shifts in climate were thus identified as 'one of the controlling causes of the rise and fall of the great nations of the world'.[25]

By 1915 Huntington turned to an even more wide-ranging discussion of the correlation between civilization and climate. Taking his examples this time from Central America as well as from Asia, Huntington repeated the claim that long-term climatic shifts accounted for the otherwise inexplicable rise and fall of civilizations like the Maya of Central America. 'Many of the great

[25]   Ellsworth Huntington, *The Pulse of Asia: a Journey in Central Asia Illustrating the Geographic Basis of History*, 2nd edn, Boston, 1919, pp. 374–81.

nations of antiquity,' he wrote, 'appear to have risen or fallen in harmony with favorable or unfavorable conditions of climate.' Like Semple, Huntington believed that only certain climates were suited to civilized life. When members of 'civilized' races strayed from their ancestral environment, they were apt to 'degenerate': this, he claimed, was what had happened to the 'poor whites' of the Bahamas. 'When the white man migrates to climates less stimulating than those of his original home, he appears to lose in both physical and mental energy.' But he was, conversely, sceptical about the benefits members of tropical races would gain by moving to more temperate climes. It would take many generations, if it were possible at all, for a changed climate to raise them to the level of the 'civilized' races.[26]

Race was given an even more prominent place in *Civilization and Climate* in 1915 than in *The Pulse of Asia* a few years earlier, with the virtues of 'Teutons' repeatedly contrasted with the alleged deficiencies of 'Negroes'. Environmentalist theory was deployed to support and legitimize pre-existing assumptions about the inherent inferiority of non-white races and their inevitable domination by whites or confinement to areas of the globe less suited to civilization. In order to try to put these claims on a 'scientific' basis, Huntington set out to map what he called human energy and creativity levels in relation to climate on a global scale. He came, not surprisingly given his original presumptions, to the firm conclusion that levels of civilization were highest in Western Europe and the north-eastern United States, but lowest in the tropics and areas of extreme cold. The ideal climate for civilization was thus shown to be (much as it had been by Montesquieu 150 years earlier) where the summers were warm, but not too hot, and where the winters were bracing but not too severe. Nations blessed with such climates possessed a high degree of will-power and energy, a capacity for progress and for ruling over others.

Although both Huntington's methods and his conclusions are viewed with great scepticism today, his work remains of interest

[26] Ellsworth Huntington, *Civilization and Climate*, New Haven, 1915, pp. 6, 33.

as an extreme, but in its time not atypical, example of how environmental determinism could be allied with racism and imperialism, and, more generally, the way in which environmental explanations have been used to support arguments of which the author was already firmly convinced. Perhaps the crudely partisan nature of Huntington's investigations into climatic influences have been of most value in persuading historians to be more wary and not to be drawn into such sweeping climatological explanations for complex processes of historical change.

The Environmental Challenge

The more historians sought to grapple with the broad themes of history and the rise and fall of civilizations, the more they felt obliged to address issues of environment and race. Arnold J. Toynbee's *A Study of History*, the first three volumes of which were published in 1934, takes us two decades further on into the twentieth century, to an age when confidence in the superiority of the West was no longer so assured as it had been and when ideas of race were taking an extreme and exceptionally ugly form in Nazi Aryanism and anti-Semitism. In a return to the grand manner of nineteenth-century historical writing and its preoccupation with civilization (though now as a species that was threatened rather than triumphant), Toynbee sought to break away from history written purely in terms of race and the rise of the nation state. He sought to go back to the cultural roots of humankind, and to establish what factors lay behind the history of civilization itself.

Toynbee felt impelled at the outset to challenge what he saw as two of the most popular explanations for the rise of civilizations. The first of these, inevitably, was race. He argued that this was a concept of relatively recent origin, which had come about in an attempt to explain the differences Europeans saw between themselves and the peoples they encountered during their voyages of discovery from the fifteenth century onwards. It had been developed in its modern form by such nineteenth-century ideologues as the Comte de Gobineau and Houston Stewart Chamberlain.

Toynbee did not believe that race as presently conceived could adequately explain the rise of civilizations, which had flourished hundreds, even thousands, of years earlier and in which there was much evidence of ethnic diversity. He could not accept that 'some special quality of race in some fraction of mankind' had somehow been responsible for the emergence of civilizations in societies so widely distributed in time and space.

But if race explained nothing, did the second popular explanation, environment? He traced this idea back to the Greeks, Hippocrates and Herodotus, and found it, by comparison with the race theory, 'more imaginative, more rational, and more humane'. But, while environmentalism was without the 'moral repulsiveness of the race-theory', it was no more convincing: indeed, it resembled the race theory in its attempt to explain the great diversity and complexity of human history from a single cause. Drawing from an even wider range of examples across time and space than Montesquieu and Huntington, Toynbee observed that if the Mayan civilization had been able to flourish in one part of the tropics, in Central America, there was no logical reason, if the environment theory were valid, why similar civilizations should not have arisen in similar areas of Africa and South-east Asia – but, he held, they had not. Two thousand years after the Mayan civilization had flourished in Mexico and Guatemala, people still lived in the forest basins of the Amazon and Congo at 'the most primitive level of savagery', while in South-east Asia the civilization of Cambodia was not indigenous but had been brought there from India. Equally, he argued, turning from hot to cold climates, Russian Orthodox civilization had not been replicated in the wilds of Canada, though it had similar environmental conditions – vast river systems, endless forests and long, harsh winters. Comparable climates, vegetations and soils did not automatically produce comparable patterns or levels of human achievement. Any convincing explanation for the genesis of civilizations had accordingly to be 'not simple, but multiple'.[27]

[27]   Arnold J. Toynbee, *A Study of History*, vol. I, London, 1934, pp. 207–44, 249–71.

And yet Toynbee found it impossible to exclude environment entirely from his discussion. He identified environmental factors among the 'challenges' that helped shape human history and drew forth the 'responses' that had helped create the civilizations of the world. The stimulus of 'hard' environments and the inventive social and technical solutions they demanded of human ingenuity were, he believed, part of a universal pattern. Taming the Yellow River's destructive floods, learning how to arrest its turbulent flow and harness its waters for rice cultivation, were, thus, among the formative influences in the creation of Chinese civilization. Likewise, in the Americas, the Maya had to invest enormous labour in clearing dense rain-forests as did the Incas and their predecessors in bringing the high Andean plateaux under cultivation. Civilization was wrested from a grudging terrain.

From such examples as these, Toynbee arrived at a general proposition about the 'stimulus of hard countries': 'the greater the ease of the environment, the weaker the stimulus towards civilization which that environment administers to Man', while, conversely, 'the stimulus towards civilization grows stronger in proportion as the environment grows more difficult.'[28] But although Toynbee saw the environment as an important illustration in early civilizations of the principle 'the greater the challenge, the greater the response',[29] he saw it as only one factor in the evolution of civilized life, and one that was of diminishing importance as history progressed. By the opening chapters of volume 3 of his twelve-volume study, with most of the ancient world behind him, Toynbee stated that civilized societies had attained increasing command over their physical environments, and his attention turned to the role of other formative factors. In 1920, in a preliminary note for his massive study, Toynbee observed:

I believe human development is a process in which human individuals are moulded less and less by their environment . . . and adapt

[28]   Ibid., vol. II, p. 31.
[29]   Ibid., vol. II, p. 259.

their environment more and more to their own will. [There came a point at which], rather suddenly, the human will takes the place of the mechanical laws of the environment as the governing factor in the relationship.[30]

There is much in Toynbee's discussion of the environment in history that today seems curiously dated, even discordant. For a start, he is unequivocally on the side of civilization, and has little sympathy for those he categorizes as 'savages'. Historical ideas about what constitutes civilization in Africa have moved on since the 1930s. Equally, although he was writing his first volumes only sixty years ago, Toynbee retained an earlier sense of how difficult and precarious was human mastery over nature: it was hard to achieve in the first place, but also difficult to maintain over any length of time. A Darwinian struggle for survival confronts all civilizations in Toynbee's analysis, and the forces of nature are among the most resilient adversaries. He rejected the view that civilization had grown up in environments that were particularly favourable for man: on the contrary (as Hippocrates had long before suggested), lands that were by nature blessed with fertile soils and natural abundance produced indolence, not civilization – as Toynbee believed to have been the case, for instance, in Nyasaland before the European takeover, or in the Amazonian jungles. The 'fallacious' view that civilization grew out of easy environments overlooked, as in the case of ancient Egypt, 'the stupendous human effort involved, not only in once transforming the prehistoric jungle-swamp of the Lower Nile Valley into the historical Land of Egypt, but also in perpetually preventing this magnificent but precarious work of men's hands from reverting to its primeval state of Nature.'[31]

Civilizations around the world had in the end succumbed to the remorseless pressures of nature, as was evident from the ruins of Anuradhapura in Ceylon, Angkor Wat in Cambodia, and the Mayan temples and palaces besieged and overrun by the jungles

---

[30]   W. H. McNeill, *Arnold J. Toynbee: a Life*, New York, 1989, p. 96.
[31]   Toynbee, *A Study of History*, vol. III, London, 1934, p. 2.

of Central America. Musing on the latter, as if on the fate of some rain-forest Ozymandias, Toynbee wrote:

> Their sole surviving monuments are the ruins of the immense and magnificently decorated public buildings which now stand, far away from any present human habitations, in the depths of the tropical forest. The forest, like some sylvan boa-constrictor, has literally swallowed them up, and is now dismembering them at its leisure: prising their fine-hewn, close-laid stones apart with its writhing roots and tendrils.[32]

But that was written in 1934. Our view of the balance between nature and culture has shifted significantly since then.

[32]   Ibid., p. 4.

3
Reappraising Nature

'La Longue Durée'

It is striking that so many of the attempts to relate history and environment have come not from historians but from philosophers, biologists and geographers. Historians have, on the whole, been more chary about invoking nature to explain the course of human history. The 'proper' study of history is still often seen to be the study of people, whether great men and women or common folk, rather than the interaction between 'man and nature'. The technical expertise of the ecologist, the climatologist and the epidemiologist have seemed to many historians irrelevant as well as inaccessible. But there have been important exceptions.

One of the most sophisticated attempts to bring together history and environment in creative partnership is to be found in the work of the *Annales* School in France, named after the journal *Annales: Economies, Sociétés, Civilisations*, in which much of their work first appeared. The journal was founded by Lucien Febvre and Marc Bloch in 1929, but the group has included Fernand Braudel and Emmanuel Le Roy Ladurie among its more recent and distinguished members.

The Annalistes have shown a consistent interest in treating the environment as a central element in human history. In part, this can be traced back to the influence of the geographer Paul Vidal de la Blache and his rejection around the time of the First World

War of crude geographical determinism in favour of a more active interdependence of people and place. Vidal's 'possibilism' was reflected in Bloch's early comment that 'counterbalancing the complexity of nature is the complexity of human emotions and human reason', and in Febvre's conclusion a few years later that geographical sites and conditions create potential lines of human activity and development but make nothing inevitable. Significantly, too, Febvre believed that it was possible to detach issues of race from the 'problem of environment'.[1] At the same time, however, the pioneers of the *Annales* School were keen to improve the status and authority of history by establishing its credentials as a science, and a more systematic and informed approach to the environment was one of the ways in which they sought to do this.

Many of these objectives can be seen in Bloch's study of *French Rural History*, published in 1931 and still highly regarded by social and economic historians. On the basis of archival research, supported by personal observation of rural France in his own day, Bloch set out to establish the factors which had shaped the physical and social characteristics of the French countryside between the breakdown of the Carolingian empire and the agrarian revolution of the eighteenth century. History, for Bloch, was 'above all the science of change', and those changes needed to be carefully dissected if the character of the period were to be accurately revealed.

In a manner that anticipated Braudel's discussion of '*la longue durée*', Bloch stressed the almost imperceptively slow evolution of life in rural France, fashioned by mutual interaction between the people and their habitat. Among the most important changes was agrarian colonization from the tenth century onwards, through the gradual winning of agricultural land from forests, marshes and heaths. In this dialectic of people and place, the forests were 'Man's most formidable obstacle', and for centuries, in medieval France as elsewhere in Europe, trees 'halted the progress of the plough'. But it was the countryside as a lived environment that

---

[1] Carole Fink, *Marc Bloch: a Life in History*, Cambridge, 1989, p. 108; Lucien Febvre, *A Geographical Introduction to History*, London, 1925, pp. 1, 235–6.

interested Bloch, not nature in the raw. The forests, like the towns, acquired their character from their human inhabitants and the trades they followed – hunters, charcoal burners, swineherds, brigands. People adapted to the landscape but at the same time brought it under their own influence and productive control. Bloch argued that 'the limitations on human activity imposed by physical environment can hardly be held wholly responsible for the basic features of our agrarian history', but they had none the less to be carefully considered when seeking to account, for instance, for the differences between France's three main 'agrarian civilizations'.[2]

The interrelated issues of time, space and culture explored in a preliminary way by Bloch in his study of French rural history were more systematically investigated by Fernand Braudel in 1949 in *The Mediterranean and the Mediterranean World in the Age of Philip II*, a work which also marked the beginning of a second, more determinist, stage in the development of the *Annales* School. What, environmentally speaking, had been implicit in Bloch and Febvre became far more explicit in Braudel. Poor Philip! The monarch who for forty years aspired to bestride the world like a colossus has to wait in the wings throughout the whole of Braudel's massive first volume and well into the second before he – and the kind of 'surface history' he is forced to represent – is allowed a proper entrance on stage. The King of Spain stands here for only one aspect of historical time, 'individual time', and for the 'traditional history' that has concerned itself with the lives of 'individual men'. This, to Braudel, is the history only of events, 'surface disturbances, crests of foam that the tides of history carry on their strong backs'. The historian who sits in Philip II's chair and reads his papers enters a 'strange one-dimensional world', 'a world of strong passions certainly', but one that is 'blind . . . and unconscious of the deeper realities of history'. This other history, with its 'deeper realities', is to be found, in part, in 'social time', represented by the activities of economies, states and societies,

[2] Marc Bloch, *French Rural History: an Essay on its Basic Characteristics*, London, 1966, pp. xxv, 5.

but more especially in what Braudel dubs 'geographical time'. This is a history built around mountains, plains, islands and isthmuses, a history of seasons, climate and epidemics; above all, it is a history of the Mediterranean, the sea which gives life and character to the whole surrounding region. Here is to be found a history 'whose passage is almost imperceptible, that of man in his relationship to the environment, a history in which all change is slow, a history of constant repetition, ever-recurring cycles'.[3]

As Braudel later explained, this history viewed over the long term and rooted in the physicality of earth, sea and sky, represented for him a 'structure', a relatively stable and enduring set of relations between people and nature. This structural element in history might find expression in other ways, such as the continuity of certain 'spiritual constraints' and 'mental frameworks'. But it was 'geographical frameworks' and 'biological realities' that Braudel principally had in mind:

> For centuries, man has been a prisoner of climate, of vegetation, of the animal population, of a particular agriculture, of a whole slowly established balance from which he cannot escape without the risk of everything's being upset. Look at the position held by the movement of flocks in the lives of mountain people, the permanence of certain sectors of maritime life, rooted in the favourable conditions wrought by particular coastal configurations, look at the way the sites of cities endure, the persistence of routes and trade, and all the amazing fixity of the geographical setting of civilizations.[4]

It is clear here, far more than in Bloch's work, that the environment gives history its underlying structure and creates a world, as one of Braudel's critics has remarked, seemingly 'unresponsive to human control'.[5] It provides the narrow physical parameters within which human beings are free to operate. There

---

[3]   Fernand Braudel, *The Mediterranean and the Mediterranean World in the Age of Philip II*, vol. I, London, 1975, pp. 20–1.

[4]   'History and the social sciences: the *longue durée*', in Fernand Braudel, *On History*, London, 1980, p. 31.

[5]   J. H. Elliott, cited in Peter Burke, *The French Historical Revolution: the Annales School, 1929–89*, Cambridge, 1990, p. 40.

seems little room here for human agency. Nature imposes limitations on what people can and cannot do, and, as becomes even more apparent from Braudel's last and most lyrical work, *The Identity of France*, it informs even the fine-grained texture of their material life: the food they eat, the clothes they wear, the houses they live in, and the character of each of the localities, the *pays*, into which France is divided.

But to see Braudel as a determinist in the Semple or Huntington mould would be mistaken and misrepresent the vast and varied nature of his work as a whole. The environment does not in Braudel simply *determine* human activity. In reviewing a book on 'human ecology' in 1944, Braudel both welcomed the author's attempt to relate climate, geography and disease to human history and cautioned against his 'biological determinism' and 'systematic reduction of the problems of man to the level of his biology'. It was not appropriate to treat human ecology as if one were dealing with the ecology of the olive or the vine, Braudel insisted: it was precisely in dealing with such matters as diet that it was necessary to address both the cultural and biological dimensions of human existence. Man 'in all his complexity – in all the density of his history, in all his social cohesion, and with all the constraints imposed by custom and prejudice' – and not just biological man – was the proper subject for historical enquiry.[6]

For Braudel the concept of the *longue durée* had a specific historical context even while being offered as a more general approach to historical interpretation and methodology. The term is essentially employed for the period of European history between the fourteenth and mid-eighteenth centuries, not far removed from Bloch's periodization in *French Rural History* and the approximate parameters of Braudel's own scholarly work. This, for Braudel, was an age when peasants constituted the great majority of the population worldwide, living from the land and tied to the 'rhythm, quality, and deficiency' of their harvests. For whole centuries, 'economic activity was dependent on demographically fragile populations.' Despite all the changes affecting

---

[6]   'Is there a geography of biological man?', in Braudel, *On History*, p. 110.

them, these four or five centuries maintained 'a certain coherence, right up to the upheavals of the eighteenth century and the industrial revolution'.[7] Like Bloch, Braudel was looking to the relative stability of the age before our own, more turbulent times, to a period when the balance between nature and humankind was both more prominent and more precarious.

Braudel's work, especially *The Mediterranean and the Mediterranean World* and his discussion of '*la longue durée*', has had a profound effect on historians working in many different fields. It has been a stimulus to attempts to integrate the environment, however defined, with other aspects of human history. His account of the Mediterranean, and its centrality within a wider geographical, social and commercial context, has prompted other historians to look afresh at different maritime and riverine systems. But in opening up the spatial and temporal parameters of history, by enlarging the environmentalist paradigm and giving it fresh credibility in historical scholarship, Braudel's work has also presented a number of problems and provoked a variety of critical responses. Is looking at the relationship between people and the environment *in the long term* necessarily the most effective and informative way of approaching it? Might there not be certain short-term episodes and events (that despised history of 'surface disturbances') which bring out even more clearly than the meandering course of *la longue durée*, and in a manner more socially and politically nuanced, the problematic nature of human interaction with the environment? And while the idea of a *longue durée*, extending over some four or five hundred years, might be of value for the Mediterranean world and France, is it equally valid for other societies – such as the Americas, for instance, in the wake of Columbus, or even in Britain in the throes of its agrarian and industrial revolutions – where the essence of environmental history might be a concern not with continuities, recurrent cycles, and slow, almost imperceptible, changes, but with radical disjunctures and irrevocable catastrophes? Conversely, for some histori-

---

[7] Fernand Braudel, *Capitalism and Material Life, 1400–1800*, London, 1973, p. 18; 'History and the social sciences', p. 32.

cally minded ecologists and anthropologists, the Braudelian *longue durée* would appear far too short to allow for the most significant continuities and changes to manifest themselves. These are issues that will recur frequently in this book. But before we attempt to address them, let us first continue to consider the contribution of the *Annales* School to the environmentalist paradigm.

Climate and History

Braudel's investigations into the relationship between humankind and the environment have been carried a stage further by several of his associates and disciples, most notably by Emmanuel Le Roy Ladurie (though it should be noted that he, like Bloch before him, has never been exclusively concerned with environmental or agrarian history). In *The Peasants of Languedoc* and *The French Peasantry, 1450–1660*, Ladurie took up from Bloch and Braudel the study of the French countryside over the long term, tracing in detail, and with the use of demographic and other statistical data, the manner in which the mass of the peasants engaged with the material world around them. The picture that emerges is one of a grim struggle barely won – a continuous fight against hunger, disease and the prospect of early death. Before the eighteenth century, France seemed physically incapable of supporting a population much above twenty million. France was already, in population and subsistence terms, a 'full world'.

In a sense, what Le Roy Ladurie, like Braudel, presents is a picture of relative stability in a human society that resembles the stasis, or stable state, that ecologists see as the mark of a mature eco-system. When France seemed set to increase beyond its 'natural' limits, it was dragged back by a series of 'ecological constraints', much as a community of herons or sticklebacks would be if they improvidently outgrew their available food supply. Death was a necessary part of this human eco-system. In the fourteenth century the Black Death savagely reduced the population of southern France, as of other parts of Europe, and reversed the laborious process of agrarian expansion which had been going on piecemeal since the tenth century. As late as the first half of the seventeenth century, rural 'reconstruction' was again checked by the triple scourge of plague, famine and warfare.

This grim view of a France largely at the mercy of hunger and disease was supported by the work of other historians, notably Pierre Goubert, in a study of the Beauvais region between 1600 and 1730 and an account of *The French Peasantry in the Seventeenth Century* published in 1982. In the work of these scholars, climatology, demography and epidemiology seem more precise tools of analysis than Braudel's expansive physical geography, and still further removed from Montesquieu's climatic speculations. In Le Roy Ladurie and Goubert, and following on from the work of the economic historian Ernest Labrousse in the 1940s, the sum of humankind's difficult and uneasy relationship with the environment is largely expressed through two linked sets of statistical data: the movement of grain prices and the rise and fall of mortality. In this attempt to quantify the human encounter with nature and to subject it to medical and meteorological scrutiny, one can see most clearly displayed the Annalistes' case for treating history as the science of the past.

And yet, powerful and potentially quantifiable though the impact of the environment is seen to be, there is a reluctance to succumb to too determinist a position. Humankind must remain more than the helpless plaything of nature: if people are not to be reduced to the status of microbes or plant colonies, they must have some conscious part in the making of their own history. In another work, *Histoire du Climat*, Le Roy Ladurie turned to one of Ellsworth Huntington's cherished themes and examined evidence of climatic change in Europe since the Middle Ages and the possible effects this might have had on economic and social conditions. Although he found considerable evidence for climatic change, he rejected the kind of determinism associated with Huntington and was highly critical of an article by a Swedish historian Gustaf Utterström which had tried to show a correlation between colder climatic conditions between the late Middle Ages and the eighteenth century and poor harvests, a declining or stagnant population, famine and high levels of epidemic mortality. 'Underlying all such theories,' scoffed Le Roy Ladurie, 'is the lazy but highly contestable postulate that climate exercises a determining influence on history.' He concluded, in rather unBraudelian fashion, that 'in the long term the human consequences of climate seem to be slight, perhaps negligible, and certainly difficult to detect.'[8]

Furthermore, in his studies of rural France, Le Roy Ladurie does not envisage the oppressive presence of hunger and disease continuing indefinitely. On the contrary (rather like Braudel and the end of the *longue durée*), by the eighteenth century the population of France was dragging itself out of its Malthusian mire and surmounting many of the ecological checks that had previously barred its progress. Human society was entering an unnatural, but arguably far happier, state. The impression one is left with, then, is that, although the environment has been an important influence on human history, and as such demands serious historical consideration, it needs to be set alongside various cultural factors if a balanced historical picture is to be achieved. Nature's hold over history may not have been broken as far back in the past as Toynbee imagined, but endured until the eve of Europe's modern age. At that point, however, in the age of the agrarian and industrial revolutions of the eighteenth and nineteenth centuries, it rapidly diminished and exercised less and less influence the closer one comes to the present day. This may be a reassuring view of the past, but is it any longer possible to regard it as an accurate view of the present?

Environmental Pessimism

As we have seen, one remarkably persistent way of understanding the past has been in terms of the power nature has exercised over human lives – shaping physical and mental characteristics, fash-

---

[8]   Emmanuel Le Roy Ladurie, *Times of Feast, Times of Famine: a History of Climate since the Year 1000*, London, 1972, pp. 24, 119.

ioning the character of laws, religions and social institutions, determining the supposed superiority or inferiority of races, governing the rise and fall of civilizations. History on the grand scale – the history of civilizations rather than the history of kings and queens – has thus often had recourse to some form of environmental determinism to supply a critical dynamic or to provide a basis for comparison between one society remote in time and space from another. But the paradigmatic status of the environment can be expressed, and in the past two hundred years has increasingly been expressed, less in terms of the guiding or controlling influence of nature over human society than of humankind's destructive, and ultimately self-destructive, dominance over nature. In order to understand this shift of emphasis, indeed virtual inversion, of the earlier environmentalist paradigm, it is necessary to refer first to theories of natural abundance and their negation.

A belief in the intrinsic abundance of nature, able to meet all human needs – food, clothing, shelter, fuel – and designed expressly by God for that purpose, is a longstanding one in Western tradition, as it has been in many other cultures as well. It is evident in the ancient notions of Paradise and Eden and their continuing appeal in religion, the arts, and one might even say in advertising and tourism as well. The idea of an earthly paradise was nurtured and sustained by European impressions of other parts of the globe, especially the tropics (as we will see in chapter 8), from the fifteenth century onwards. It was hard for Columbus, for instance, not to believe that he had stumbled upon the Garden of Eden when he arrived in the West Indies in 1492. Similarly, when reports of the Pacific island of Tahiti began to reach Europe in the second half of the eighteenth century, it seemed to many that paradise had been rediscovered in an island setting and among a people free (as Europe demonstrably was not) from hunger, misery and disease. 'I thought I had been transported into the Garden of Eden,' remarked the French navigator Louis Antoine de Bougainville on his arrival at Tahiti in 1768. His companion, Philibert Commerson, was no less ecstatic, declaring that Tahiti was a veritable Utopia:

inhabited by men without either vices, prejudices, wants or dissensions. Born under the liveliest of skies, they are supported by the fruits of a soil so fertile that cultivation is scarcely required and they are governed rather by a sort of family rather than by a monarch . . . they recognize no other god save Love. Every day is consecrated to him and the whole island is his temple . . .[9]

But the idea of natural abundance in the tropics seemed to stand in stark contrast to the parsimony of nature in Europe, or to be simply a figment of the sea-weary traveller's imagination. In his *Essay on the Principle of Population*, first published in 1798, but extensively revised in subsequent editions, T. R. Malthus took a far more pessimistic view of nature and the limits it imposed. He identified what he saw to be a universal tendency of human beings (shared, significantly, by plant and animal populations) to multiply exponentially until constrained by the available supply of food. The inevitable consequence – so inevitable in his view that it constituted a law of nature – was a series of checks, principally in the form of disease, war and famine, which forced the population back to the size the limited food supply could sustain. Human societies might sensibly anticipate the onset of these 'positive' checks by having recourse to 'preventive' strategies of their own. Some of these – such as infanticide – were the mark of the savage and had no place in civilized countries, where moral and sexual restraint through celibacy and late marriage would alone limit the number of births and hence the number of mouths to feed. A clergyman by training, Malthus believed that this was God's way of encouraging industry and restraint. Only by working hard rather than following their slothful inclinations, only by restraining the passions that created more children than parents could possibly feed, was it possible for people to survive and fulfil divine expectations.

In Malthus, nature had a dual role. First, it stood for plenitude and fecundity, the tendency of all living creatures – animals,

---

[9]   Quoted in Richard H. Grove, *Green Imperialism: Colonial Expansion, Tropical Island Edens and the Origins of Environmentalism, 1600–1860*, Cambridge, 1985, p. 238.

plants, people – to reproduce beyond their means. But, secondly, it stood for the destructive forces – famine, pestilence and disease – which are constantly at work to contain this profligacy.

Malthus was writing mainly with his own society in mind and the need to restrain what he saw as the growth of an ever-expanding class of unemployed poor, who depended on charity but who none the less continued to breed beyond their means of support. But what is striking, especially in the later editions of his essay, is the extent to which Malthus drew, much as Montesquieu had done fifty years earlier, upon a wide range of examples taken not just from the classical world of Greece and Rome and from contemporary Europe but also from recent reports of life among the indigenous peoples of Africa, Asia and the Americas. But, in contrast to Montesquieu, whom he occasionally and disapprovingly cited, Malthus attached little importance to climate and to other environmental factors. Population, not climate, was in his view the true determinant of history: what demarcated civilization from savagery was not the presence or absence of a suitably stimulating climate but the ability to regulate population in line with the capacity to produce food. Malthus rejected climatic determinism just as he scorned ideas of natural abundance. Tahiti, by his account, far from being an earthly Eden, was ravaged by warfare, hunger and disease, and, the victim of an improvident attachment to the god of Love, could only control its numbers through the savage checks of infanticide and human sacrifice.[10]

Malthus's legacy has proved remarkably enduring: 'few men in the history of Western thought have exerted an influence comparable to his.'[11] His essay had a profound impact on two other founding fathers of modern environmentalism, Humboldt and Darwin. Worster describes Darwin's reading of Malthus in 1838 as 'the single most important event in the history of Anglo-

[10] T. R. Malthus, *An Essay on the Principle of Population*, 1st edn 1798, London, 1973, pp. 44–58.

[11] Clarence J. Glacken, *Traces on the Rhodian Shore: Nature and Culture in Western Thought from Ancient Times to the End of the Eighteenth Century*, Berkeley, 1967, p. 637.

American ecological thought'.[12] And Darwin himself described the struggle for survival in nature as

> the doctrine of Malthus applied with manifold force to the whole animal and vegetable kingdoms; for in this case there can be no artificial increase of food, and no prudential restraint from marriage. Although some species may be now increasing, more or less rapidly, in numbers, all cannot do so, for the world would not hold them.[13]

How far Darwin was actually influenced by Malthus remains a moot point. Perhaps he merely found in Malthus timely confirmation for an idea of competition for finite resources he had already begun to develop as a result of observations made during the voyage of the *Beagle* a few years earlier. But, certainly, the idea of the Earth being unable to sustain exponential increases in human or animal populations, as first presented by Malthus and then endorsed by Darwin, has had an immensely powerful effect upon subsequent environmentalist thought. If one part of ecology has emphasized balance and harmony within nature, another has seen competition between and within species for finite resources as a central mechanism in controlling populations and forcing a choice between adaptation and extinction.

Malthus himself had scant interest in the potentialities of human modification of the environment and little faith in the capacity of agriculture to keep pace with demographic growth – curiously, it might seem, for someone who lived at a time when the English landscape was undergoing an unprecedented transformation with the agricultural and industrial revolutions. But others within a generation or so of Malthus were becoming increasingly conscious of man-made environmental change and its effects. Perhaps nowhere was the 'modification of nature by human action' more visible than in the United States where what had once been vast forests had within a few years been transformed

---

[12]   Donald Worster, *Nature's Economy: a History of Ecological Ideas*, Cambridge, 1985, p. 149.

[13]   Charles Darwin, *On the Origin of Species*, 1st edn 1859, Harmondsworth, 1968, p. 117.

into cleared farmland and thriving settlements. Such changes in Europe had taken centuries; in America, they happened almost overnight. George Perkins Marsh in 1864 was one of the first American writers to put these concerns into print. To many American environmentalists, Marsh has become an ecological hero, the pioneer of what has since become the conservation movement, the harbinger of a profound shift in Western attitudes to nature.

But in his book *Man and Nature* Marsh was less concerned about the fate of plants, animals and other living occupants of the planet than with the destiny of its human inhabitants. He used historical as well as contemporary examples to illustrate his basic contention that 'the earth is fast becoming an unfit home for its noblest inhabitant', by which, clearly, he meant humankind. Where many writers had believed that 'the earth made man', Marsh was of the opinion that 'man in fact made the earth', and his tampering and interference now threatened to destroy nature and with it humans themselves.[14] *Man and Nature* was not only geographically wide-ranging; it also covered an impressively large range of subjects. The subject headings of his chapters read almost like agenda items for a present-day environmental conference: 'Transfer, modification and extirpation of vegetable and animal species', 'The woods', 'The waters', 'The sands'. And, not least important, for a man who had witnessed the effects of tree-felling in his native Vermont, was the prominence he gave to forests and deforestation in environmental destruction and decay. Marsh was certainly not alone in this – the impact of deforestation on climate and soil had been a matter of growing scientific concern for two hundred years – but *Man and Nature* helped mark the entry into modern environmentalist debates of trees as both a symbol of nature and a living link in maintaining an equable balance between nature and humankind.

---

[14]   George Perkins Marsh, *Man and Nature: or Physical Geography as Modified by Human Action*, lst published 1864, Cambridge, Mass., 1965, p. ix.

A 'Green' History

Since Marsh, environmentalism has spawned many different types of historical approach and enquiry. But, as a concluding example of environmental pessimism, let us turn, finally, to Clive Ponting's *Green History of the World*. The contrast with earlier writers is striking. Unlike Toynbee or Huntington, this is not a history centred around the rise and fall of civilizations. It is a blistering account of how the planet has suffered and sickened under human tutelage. It is neither climate nor disease that represents the environment but all forms of nature, animate and inanimate combined, and the hazards to which they have been exposed in a world increasingly misruled by humankind.

Malthus, more than any other founding father of environmentalist thought, sets Ponting's agenda. A history of humans' inherent and ultimately senseless profligacy is set against the limits of finite space and limited resources. The Earth, as with Malthus, is a 'closed room': it is in danger of becoming a tomb as well. Ponting explains how, in order to feed their growing numbers, humans moved from hunting and gathering around 10,000 BC to settled agriculture. This had momentous and often destructive consequences for the environment, as was evident from the fate of several early civilizations. But the problem of feeding an ever-growing population remained. Further innovations were achieved to overcome this, but these, too, had negative effects. The overseas expansion of Europe became the 'rape of the world', with the wholesale destruction of native wildlife, the introduction of alien species, and the creation of a resource-reckless global economy. A second 'great transition', greater even than the neolithic revolution, followed the first with the use of fossil fuels and the spread of industrialization. Yet, despite these expedients, and in part because of them, human numbers increased ever more rapidly. It took two million years, Ponting notes, for the world's population to reach one billion (in 1825), but only another century to reach two billion and a further thirty-five years to reach three billion (in 1960). The next billion was added in only fifteen years and five billion came, in even shorter time, by the late 1980s.[15]

The rise of the world's population since the First World War is undeniably a major factor in any consideration of environmental issues today. But how far is it meaningful to project today's Malthusian concerns back into earlier history? Ponting (like his mentor, Malthus) takes population to be the crucial factor throughout all human history and reads its effects and its significance far back into the past. The trail of destruction begins, not, as one might imagine, with industrialization and nineteenth- and twentieth-century urbanization, but is a remorseless process going back 10,000 years to the very origins of settled societies. What to Toynbee was the heroic wresting of civilization from 'hard' environments and resilient nature is for Ponting the irreversible destruction of fragile eco-systems from ancient Mesopotamia and the Indus Valley onwards.

Given such a presentist approach to history, it is not surprising that Ponting rejects the old linkage, still alive in Toynbee or Huntington, between civilization and human mastery over nature, and instead sees almost any kind of human activity as a potentially fatal assault on the precarious balance of natural eco-systems. Humankind stands so far apart from nature that its very existence is inherently a threat to the survival of any natural eco-system. This, moreover, is a global history, one in which the entire Earth shares a common destiny. What to Hippocrates, in the fifth century BC, or in the neo-Hippocratic thought of the eighteenth century, was the malign influence of some airs, waters and places, has become on the eroded, polluted, overcrowded planet of the late twentieth century the toxicity of virtually all airs, waters and places.

Ponting begins, as in a rhetorical sense he leaves us, with the 'lessons' of Easter Island. Islands, especially tropical islands, have (as in the case of Tahiti) long had an important emotional and intellectual niche in environmentalist thought, for the uniqueness and fragility of their isolated eco-systems and for their symbolic

---

[15]   Clive Ponting, *A Green History of the World*, Harmondsworth, 1992, p. 240.

significance as 'closed rooms' with finite space and resources. When in the eighteenth century European navigators first visited Easter Island, divided by more than a thousand miles of Pacific Ocean from the nearest inhabited land, they were amazed and mystified by the huge stone statues they found there. It seemed inconceivable that three thousand 'primitive' inhabitants living in reed huts and caves, on an island only 150 miles square and almost entirely devoid of trees, could have been responsible for carving and erecting over six hundred statues, many over twenty feet high. It seemed more probable that some outside civilization had been responsible.

In fact, Ponting explains, the islanders did make the statues themselves, but in felling large numbers of trees in order to move the statues from the quarries to the erection sites they stripped the island of its trees and other natural vegetation. A population estimated at seven thousand at its peak in the 1550s shrank rapidly and the islanders were unable to sustain their former level of civilization. Having nowhere else to go for food and other resources, they lapsed into the 'near barbarism' in which the Europeans found them.

The Earth, too, Ponting warns, is a 'closed system', threatened with a similar 'self-inflicted, environmental collapse'. 'Green history' is a history of human folly and destruction: for the fate of Easter Island read Planet Earth. But, as we have seen, historians even in the recent past have not always portrayed the relationship between nature and culture in such bleak, dichotomized and unrelentingly Malthusian terms. For many, the relationship is not only more complex, but also far more ambivalent.

4
Environment as Catastrophe

Continuity and Crisis

In history, as in ecology, there is an abiding tension between stability and change. Ecologists tend to favour the idea of nature as a self-regulating and stable system – so long as there is no external interference to disrupt the balance. Alexander von Humboldt, the German naturalist and a pioneer of modern ecological thought, captured this view of nature when he described the cosmos as a 'harmoniously ordered whole', 'a harmony, blending together all created things, however dissimilar in form and attributes; one great whole animated by the breadth of life'.[1] But as Humboldt recognized from the geological record, and as Darwin saw in South America and elaborated in his Malthusian account of the 'struggle for existence', not all in nature was harmony and stability. Volcanoes and earthquakes wrought violent changes in the face of the earth; seas and mountain ranges rose and fell; species were driven to extinction through competition or the destruction of their habitat.

Many recent ecologists see a place for both slow evolutionary change and for more rapid, even revolutionary, transformations. James Lovelock in his controversial Gaia thesis which represents

[1] Alexander von Humboldt, *Cosmos: a Sketch of a Physical Description of the Universe*, vol. I, first published 1845, London, 1901, pp. 2–3.

the Earth as a single, complex, living organism, has argued that 'from the long Gaian view the evolution of the environment is characterized by periods of stasis punctuated by periods of abrupt and sudden change.' The environment was, he believes, never so uncomfortable as to threaten the complete extinction of life on Earth, but during periods of accelerated change 'the resident species suffered catastrophe whose scale was such as to make a total nuclear war seem, by comparison, as trivial as a summer breeze is to a hurricane.'[2]

As it is with ecology, so has it been with history. Some historians have been drawn to the revolutionary moment and the bitterest human struggles for survival; others have found more meaning in the slow-moving evolution of human ideas and institutions. And there have been corresponding differences of approach in historians' views of nature and the ways it impacts upon human societies. One historian represents the effect of nature as gradualist or cyclical; another sees it as catastrophic. In the former, nature tends to be a background force or an increasingly accommodating presence. It operates through the recurrent cycle of the seasons, or the emergence of a broad equilibrium between people and their environment. Society is influenced as much by local conditions as by more general forces of change. Climate, soil, vegetation, domesticated animals, disease and human land-use combine to produce a distinctive physical landscape and social and cultural forms appropriate to it. This is the view of environmental history that informs much of the writing of the *Annales* School, especially Fernand Braudel's work on the Mediterranean and France. It is also, in softer hues and less scholarly tones, the mellow understanding of nature and its harmonious relations with human life to be found in Gilbert White's *Natural History of Selbourne*, first published, paradoxically, in that year of revolution, 1789.

The second view of nature might accept the long-term significance of environmental factors in quietly shaping the general character of human history, but it is not content to stop there. It

[2] James Lovelock, *The Ages of Gaia: a Biography of our Living Earth*, Oxford, 1989, pp. 153–4.

sees the even flow and cyclical rhythms of history as being massively interrupted from time to time by major environmental crises and disjunctures. The forces of nature for once step out of the shadows to play a direct and temporarily decisive role in human affairs. Catastrophe strikes, and the old order is overturned, gone perhaps for ever. The ecological event may, in itself, be shortlived, but its impact reverberates for generations. In exceptional cases the whole history of a region, even an entire continent, is diverted along new paths, or emergent trends receive a new and unexpected impetus (though it remains a problem for the historian to decide how far it was the ecological event which caused such far-reaching changes or whether it merely provided an opportunity for their emergence).

Reflected in these approaches to nature are two more general differences of attitude to history – the history of a society epitomized or transformed by moments of crisis rather than by the slow and faltering evolution of ideas, institutions and practices. In an age in which rapid environmental change and incipient crisis is apparently all about us – in the destruction of tropical rain-forests, depletion of the ozone layer, the onset of global warming – a cataclysmic approach to the relationship between human history and the physical environment might well appear more relevant than a slow-moving history that gradually unfolds over the vast expanse of *la longue durée*. But there are dangers in the appeal of the catastrophic. Biblical images of Noah's flood, or the plagues of Egypt, readily spring to mind. The Black Death becomes the archetype for every other epidemic, real or imagined. Environmental crises, now as in the past, are understood in apocalyptic terms or become the vehicle for crudely Malthusian or Darwinian assumptions about how societies function as they teeter on the edge of an ecological abyss. While pleading the objective authority of science, we invest the environment and its history with stark moral messages: an epidemic or an earthquake becomes a construction of our own creation, a vehicle for our subjectivity.

Environmental history (again, like ecology) also raises complex issues of scale. While it is possible to focus upon the microcosmic relations between people and nature within a single locality and over a short span of time, more often environmental historians are tempted to speculate on a far larger scale, making it all the more difficult to determine cause and effect and to avoid drawing sweeping conclusions from scanty evidence. On the parochial scale of the quiet Hampshire village of Selbourne, the 'great storm' of 1703 (seventeen years before Gilbert White's birth) was a calamity of sorts, an event worthy of recall sixty years later. The 'amazing tempest' tore down a 'vast oak', a 'venerable tree' which had been 'the delight of old and young, and a place of much resort in summer evenings'. Doubtless, too, until its destruction the tree was equally welcomed as a place of resort for many of Selbourne's birds and insects.[3] On a far larger human scale, the Lisbon earthquake of 1755 not only caused some 40,000 deaths, but, as evidence of nature's unpredictability and destructive force, sent cultural shockwaves throughout Europe, as Voltaire's *Candide* attests. But the catastrophes with which environmental history now commonly concerns itself are often of even greater magnitude – appealing to some historians' search for the origins and agencies of change on a continental, or even global, scale.

Earthquakes, floods and volcanic eruptions have been identified as having profound effects on human society – disrupting climate and jeopardizing cultivation over vast areas, dealing the final blow to ailing civilizations, or, as much by their unexpectedness as by their magnitude, sending shockwaves far beyond the region immediately affected. There are many examples of this search for grand ecological causes in recent historical literature. John D. Post has argued that in the pre-industrial age European societies were particularly vulnerable to sudden climatic fluctuations of the kind that resulted from the huge dust clouds thrown up by the volcanic eruption of Mount Tambora in the Indonesian archipelago in April 1815. The subsequent phase of cold, wet weather is said to have resulted in three years of poor harvests, hunger and social disruption in Europe and other parts of the world. While

[3] Gilbert White, *The Natural History of Selbourne*, 1st edn 1789, London, 1887, pp. 6–7.

wary of Huntington's 'facile assumptions' about a direct correlation between climatic change and human history, Post none the less argues from this example that

> certain crises can be shown to have resulted from adverse meteorological patterns. The dynamic nexus is not difficult to elaborate: adverse weather, deficient crop yields, economic distress, malnutrition or famine, fewer marriages, a reduced birth-rate, an increased death-rate, widespread begging, mass vagrancy, and epidemic disease.[4]

But how far such dramatic and relatively shortlived episodes actually force significant and lasting changes in, for example, agricultural practices or political life remains doubtful. As one of Post's critics rather caustically remarks, to focus on such 'unique, exogenous shocks', while failing to quantify what proportion of contemporary misfortunes was directly attributable to such events, is like writing a history of banking solely in terms of bank robberies.[5]

Historians of climate have often felt more confident in looking at long-term shifts rather than at short, sharp shocks. Climatic change has been credited, for instance, with having had a momentous effect on Europe in the late Middle Ages. For several centuries northern and western Europe had benefited from a relatively warm climate and favourable agricultural conditions. But, by the early fourteenth century, temperatures had begun to fall, on average one or two degrees centigrade below present-day levels, enough to shorten the growing season significantly and make bleaker upland areas unsuitable for cereal crops. The effects of this climatic shift, which intensified into the 'unpleasant meteorology' of the Little Ice Age in the sixteenth and seventeenth

---

[4] John D. Post, 'Meteorological historiography', *Journal of Interdisciplinary History*, 3, 1973, p. 725; see also *The Last Great Subsistence Crisis in the Western World*, Baltimore, 1977.

[5] Jan de Vries, 'Measuring the impact of climate on history: the search for appropriate methodologies', in Robert I. Rotberg and Theodore K. Rabb (eds), *Climate and History: Studies in Interdisciplinary History*, Princeton, 1981, p. 23.

centuries, has been the subject of much recent historical debate. Possible effects include diminished harvests, increased epidemic mortality, and the curtailment of Norse voyaging across the north Atlantic to Iceland, Greenland and Newfoundland. A colder climate and the southward spread of polar drift-ice contributed to the abandonment of Greenland in the fourteenth century and the decline of Icelandic agriculture. This closing of the North Atlantic frontier also ensured that when Columbus pioneered a new sea route west in 1492 it was in warmer latitudes and in Iberian caravels, not Norwegian longboats.

But while climate – especially sudden climatic change of the kind Post identifies – continues to be discussed as one possible factor influencing human history on a vast human and geographical scale, it is more commonly through the eruption of epidemic diseases rather than of remote volcanoes that scholars have sought most dramatically to represent nature and its historical impact. Enough epidemics and pandemics have marked the past two thousand years of history for there to be several relatively well-documented examples of their capacity for destruction and dislocation on a huge scale. Their effect has often been so immediate and yet so far-reaching, that their historical impact has been impossible to deny, even if historians have struggled to find the appropriate means to understand their causes and evaluate their effects. In becoming more environmentally aware, historians have been forced to become more technically minded, drawing freely upon climatology, epidemiology and ecological science. But, as with the historical assessment of other natural phenomena, it remains difficult to evaluate the complex and often contradictory data about epidemic disease, and to establish a balance between what is 'natural', on the one hand, and cultural or 'man-made', on the other. Some historians are drawn by their technical expertise to give precedence to nature over culture, while others, anxious to preserve a human focus, stress the importance of human experience and response. The question thus remains: are even the greatest of 'natural disasters' in the end culturally rather than environmentally determined?

Europe's 'Greatest Environmental Crisis'

It has been argued that the greatest single environmental crisis to hit Europe (as well as much of the Middle East and North Africa) in the past two thousand years was the Black Death, the epidemic of bubonic plague that between 1346 and 1351 blazed a trail of mortality from the Black Sea to the Baltic and from Egypt as far as Iceland. Some twenty million people are estimated to have perished in Europe alone. This, according to Robert S. Gottfried, was Europe's 'greatest ecological upheaval': 'The effects of this natural and human disaster changed Europe profoundly, perhaps more so than any other series of events. For this reason, alone, the Black Death should be ranked as the greatest biological –environmental event in history, and one of the major turning points of Western Civilization.'[6]

The Black Death was not the first major epidemic, even of bubonic plague, to have devastated Europe. In 430 BC Athens experienced a 'plague' caused by an unknown disease, while epidemics of what are thought to have been smallpox and measles ravaged the later Roman Empire. In 541–2 Constantinople and the eastern Mediterranean suffered 'the plague of Justinian', an outbreak of bubonic plague that appears to have originated in East Africa and the Upper Nile. This brought a mortality estimated at 20–25 per cent to southern and eastern Europe, and lingered on in parts of the Mediterranean basin for the next two hundred years, causing much death and desolation, without ever becoming wholly endemic.

But the Black Death brought plague to Europe on an unprecedented scale, and caused mortality exceptional in the history of any disease. It is estimated to have killed at least a third of the population and in some areas significantly more. Although it took six years for the epidemic to trawl through Europe from end to end, locally most of the mortality occurred within a few months.

[6] Robert S. Gottfried, *The Black Death: Natural and Human Disaster in Medieval Europe*, London, 1984, p. 163.

Ports, the larger towns and cities, and the more densely populated and commercially active parts of the countryside suffered most. Italy felt the impact of the epidemic first. Plague robbed the port of Genoa of between 30 and 40 per cent of its 100,000 inhabitants. Sienna lost perhaps half of its 60,000 people, and Florence a third or more of its citizens in six months. Across Italy as a whole, as many as 40 per cent of the population may have died. The south of France suffered heavily, too, with Languedoc, one of the wealthiest provinces, estimated to have lost nearly half its inhabitants. Across southern Europe between 35 and 40 per cent of the population perished.

Mortality of a third or more was also recorded in parts of northern Europe – in Normandy, the Low Countries and Scandinavia. In England, where plague mortality has been more closely scrutinized than elsewhere, as many as 40 per cent of London's 50,000 people may have died, and, Bristol, England's second largest city, suffered a similar scale of loss. Rural East Anglia was likewise severely affected, with mortality rates close to 50 per cent. By contrast, some parts of Europe, especially in the centre and east, escaped relatively lightly – notably Poland, where the Black Death arrived late and caused a mortality of less than 25 per cent. But across Europe as a whole the demographic impact was devastating: one-third of Europe's 75 million people had been wiped out in five years. 'So many died,' lamented Sienna's chronicler Agnolo di Tura of his afflicted town, 'that all believed it was the end of the world.'[7] He might have been speaking of the fate of Europe as a whole.

Even so there was no lasting respite. The Black Death of 1346–51 was just the opening salvo in the protracted history of the second plague pandemic. In 1361 bubonic plague returned to Europe a second time, carrying off many of the young born since the previous epidemic. It struck again in 1369, killing a further 10–20 per cent of Europe's population. Thereafter, and until well into the sixteenth century, plague returned on average every ten to twelve years to haunt Europe and to carry off its deadly tithe.

[7]   *Cronica Senese di Agnolo di Tura*, cited in ibid., p. 45.

Not until the seventeenth century did its grip weaken: the Great Plague of London in 1665 was one of the last major outbreaks in north-western Europe, though Marseilles suffered a serious epidemic as late as 1720. It has been argued that it was not so much the first devastating epidemic of the 1340s which made plague such a momentous force in European history, but these repeated hammer blows, which kept the population of Europe below pre-plague levels for generations. At the same time, plague moved from being a single epidemiological and demographic event to become a cyclical phenomenon, a recurrent factor, along with war and famine, in the 'endemic disequilibrium' that characterized Europe's demographic *ancien régime* until the late eighteenth century.

The effects of the Black Death and its epidemic aftermath have been widely documented and discussed. Only a brief sketch is necessary here to indicate the enormous impact that has been attributed to this 'the most cataclysmic of all epidemic diseases'.[8] In the European countryside a process of agricultural expansion had been going on for three or four centuries, as Marc Bloch showed in the case of France, turning forests, moors, heaths and fens into agricultural land and bringing even marginally productive land under the plough. The coming of the Black Death and the resulting rural depopulation, however, put 'a decisive end to the first great wave of medieval colonization'.[9] Villages were abandoned, many of them for ever, and erstwhile arable land was given over to grazing sheep. A disease so destructive of human life, plague at least brought some respite to Europe's wildlife (including its dwindling population of wolves), and allowed the fast vanishing forests and wetlands a chance to survive and recover.

On the social and political front, too, plague's impact was momentous. In Britain in particular, plague signalled the end of

[8]   Andrew A. Appleby, 'Epidemics and famine in the Little Ice Age', in Rotberg and Rabb (eds), *Climate and History*, p. 67.
[9]   W. G. Hoskins, *The Making of the English Landscape*, 1st edn 1955, Harmondsworth, 1985, p. 85.

feudalism, accelerating the change from a society founded on personal service to one based on a money economy. In a situation where land was plentiful but labour scarce, the manorial system decayed, allowing serfs to become rich peasants and, in time, yeoman farmers. Those of the lower classes who survived the repeated plague epidemics enjoyed a higher standard of living than previously. By the middle of the fifteenth century, real wages in England were two and a half times higher than in the early thirteenth century, and may have been higher in the fifteenth century than at any subsequent time before the twentieth century. Plague helped turn the late Middle Ages into the labourers' 'golden age'.

Faced with declining incomes and labour shortages, rural and urban élites struggled to restore pre-plague wage levels and working practices. In England, Parliament introduced first the Ordinance (1347) and then the Statute of Labourers (1351) in an attempt to impose what would now be called a 'statutory wage-freeze'. But such measures generated their own reaction and sparked a wave of popular discontent and revolt, typified by the revolt of the Ciompi labourers and artisans in Florence in 1378, the peasant Jacquerie in France in 1358, and the English Peasants' Revolt of 1381. Plague also triggered far-reaching changes in agriculture, manufacturing and trade. Before the Black Death, production of agricultural and industrial goods had relied on the availability of cheap and abundant labour. After it, there was greater reliance upon more sophisticated technology and a more cautious use of the scant supply of labour: this, arguably, was one of the ways in which Europe was beginning to rely more heavily on technological innovation and diverge from the populous societies of eastern and southern Asia.

Religious life, too, was affected. Not only did the Church lose a large percentage of its priests and scholars, but discontent grew over the failure of the clergy to do more to provide help and comfort during the plague years. In a time of deep spiritual crisis, popular religious ideas took a morbid, millennarian and heretical turn, epitomized by the Flagellants, who wandered through towns and villages, whipping themselves and preaching repentance to ward off the divine punishment of plague. Some of the seeds of doubt and dissatisfaction with the established Church sown by plague in the fourteenth century were to eventuate in the Reformation one hundred and fifty years later. Popular attitudes also took other forms with equally lasting effects. In a score of German towns and principalities, as in southern France and Spain, Jews were blamed for the outbreak of plague. Dozens of long-established Jewish communities were destroyed, and thousands of Jews fled eastwards to Poland and Russia, thereby bringing about a significant geographical relocation of the Jewish population and redefining their role in Europe's history. In art and literature, too, death and suffering shut out happier sentiments and obscured more tolerant attitudes: 'darker climates of opinion and feeling' came to prevail.[10]

Historians have further argued that the Black Death was a factor in changing Europe's relations with its neighbours, directing it onto a new and momentous course. In combination with the climatic changes mentioned earlier, plague helped sever links with Greenland, the last ship leaving from the disease-stricken port of Bergen in 1369. In the Iberian peninsula, as in Scandinavia, the Black Death closed down one phase of European expansionism: much of what had been chanced upon or discovered in the northern Atlantic was forgotten or relegated to myth. Conversely, and it is a point we will return to in the next chapter, the shortage of labour, caused by the loss of a third of the population prompted an increased demand for slaves in southern Europe. This could no longer be met, as formerly, from the Black Sea region, which had itself been severely depopulated by plague and was further cut off from European traders by the fall of Constantinople to the Ottoman Turks in 1453. Italian merchants and Portuguese adventurers looked increasingly to Africa as an alternative source of slaves, thus giving impetus to slave-raiding and exploration further and further down the West African coast. If in some respects plague bred introspection and doubt, the

[10] William H. McNeill, *Plagues and Peoples*, 1st edn 1976, London, 1979, p. 159.

'great age of death' also prompted Europe to look outwards in search of slaves, gold and spices.

Plague also stimulated a fresh approach to medical observation and enquiry and encouraged the introduction of new sanitary measures, such as the use of quarantines at ports and the isolation of infected households. In the long term, these anti-contagious practices may have helped to limit Europe's exposure to a disease which lingered on in the Middle East into the nineteenth century, long after its disappearance in the West. Even if such sanitary measures yielded few unambiguous results, they produced enough success locally to promote confidence in the belief that human action could exercise some controlling influence over disease, that humans could counter the inroads of even so formidable an adversary.

Thus, across a broad spectrum of social, economic and political life plague left an indelible mark on late medieval and early modern Europe – or so it is widely thought. The precise form of human responses to plague necessarily depended upon pre-existing social and economic conditions (such as the weaknesses of the manorial system or latent anti-Semitism), and might vary from one culture to another (Islamic attitudes to plague did not mirror Christian responses, nor was there a uniform reaction even within those religious traditions). None the less, plague is widely credited with having initiated or accelerated wholesale change in Europe. If, in Toynbee's terms, the Black Death was a major environmental 'challenge' to European civilization, then the capacity to survive it and to turn the experience to constructive use, must surely be reckoned one of the most formative episodes in the rise of modern Europe.

On the face of it, then, the Black Death appears to offer a massive – even archetypal – illustration of how the course of history can be changed not by individual will, or by political acts and ideologies, but by environmental forces – even if, as in this instance, the disease is normally no more than incidental to humans and hence its impact on history is in many respects purely fortuitous. The point is worth pursuing. Plague is caused by a bacillus, *Yersinia pestis*, a micro-organism that occurs among several different rodent species. It is transmitted by blood-sucking fleas, normally rat fleas, but possibly in certain circumstances by human fleas as well. The bacilli multiply in the flea's stomach until it cannot ingest blood without regurgitating large numbers of them into its victim. The bacilli are able to survive for long periods among rodent communities, and without becoming a hazard to their human neighbours. But periodically, for reasons which are not yet fully understood, the disease spreads from its normal rat hosts to affect other rodents less tolerant of the infection, causing mass mortality among them. As their rodent hosts die, the fleas turn from rats to people, carrying the plague bacillus with them in their bite. Thus plague as a human disease is the result of a transference to humans of a pathogen customary only among rodents. It is a disease dependent upon an ecologically complex combination of bacilli, rats, fleas and humans, aided by the right climatic conditions.

The character as well as the timing of epidemics are also closely related to the ecology of the plague bacillus. In humans, plague takes three main forms – bubonic, pneumonic and septicaemic. Of these, the bubonic form was historically the most common. About six days after infection by the flea's bite, the site affected developed a blackish patch under the skin. The lymph nodes in the groin, armpit, or neck began to swell with dead tissue and cell debris, forming the characteristic 'bubos' or swellings that gave the disease its name. Of those thus infected, 60 per cent were likely to die, usually within about ten days. Although this was probably the most common form of the fourteenth-century plague, the two other forms were also present, contributing to the high levels of mortality recorded and the speed with which the disease spread. In the most infectious form, pneumonic plague, the bacillus spread directly from person to person by airborne transmission rather than through fleas, while in septicaemic plague the bacilli entered the bloodstream directly, multiplying massively within the body's organs and causing rapid death. In these latter forms of plague, mortality was often in excess of 95 per cent of those infected.

One of the characteristics of the second plague pandemic

appears to have been the presence of the pneumonic and septicaemic forms, suggesting that person-to-person contact or transmission via human fleas rather than the prior infection of resident
rodent communities may explain the remarkably rapid spread of
the disease across Europe. But one of the problems of disease
history is that diseases do not necessarily remain the same over
time or function identically in different localities. The way in
which bubonic plague spread in the fourteenth century suggests a
rather different pattern of incidence and transmission from that
in the more closely observed third pandemic that began in east
Asia in the 1890s.

Plague has had no permanent home in Europe. Historically, its
reservoirs of infection have been elsewhere – in Central Asia,
Siberia, south-western China, the Middle East and East Africa.
However, during the fourteenth and fifteenth centuries the disease
lodged among the rodents of Europe, hence the recurrent epidemics of the period. It has been something of a historical 'whodunit'
to establish when and how bubonic plague broke out of its
ecological isolation to cause such murder and mayhem across
Asia and Europe.

The pan-Asian empire of the Mongols, stretching from China
to India and the eastern borderlands of Europe, has been seen as
providing plague with a unique opportunity to set out on its wider
peregrinations. At its military and political zenith between 1279
and 1350, the Mongol empire facilitated trade between East and
West, across the steppe as well as by the older, more southerly,
Silk Route. It allowed armed horsemen to move rapidly from one
hitherto secluded disease environment to another, carrying pathogens and their obliging vectors along with them in their booty
and saddlebags. Pursuing a more general thesis about the role of
the Asian steppes in Eurasian history, W. H. McNeill has argued
that Mongol soldiers, penetrating south into the Chinese province
of Yunnan and the neighbouring regions of Burma in the mid-
thirteenth century, entered an area where the plague bacillus was
already endemic. From there the disease was brought back to
Mongol headquarters at Karakorum in Mongolia, with grain or
cloth harbouring infected fleas, and in this way the rodents of the

steppe became infected. From this strategic reservoir of infection, plague set out, according to McNeill, on an expansionist career even more far-reaching than that of the Mongols themselves. First, it advanced into China (where there is known to have been massive mortality from the 1330s onwards, though from what cause is unclear); then into India and the Black Sea region by 1346. From there, given the extent of Italian trade with the Crimea and Levant, the spread of the disease into southern Europe was swift and easy.[11]

McNeill's thesis remains in several respects highly speculative. Did plague need to be imported from Yunnan, or was it already endemic among the rodents of Central Asia before the fourteenth century? Did climatic change, rather than human agency, cause rodents to move from their customary habitat, thus increasing human exposure? Was plague the cause of mortality in China in the 1330s and 1340s, or was it some other disease? Did plague reach the Black Sea from Central Asia or erupt from some nearer focus of infection in India or the Middle East? None the less, although the origins of the plague pandemic remain obscure, there can be no doubting the importance of cultural factors, broadly defined, in the spread of plague once it reached Europe. The volume and direction of maritime traffic in the Middle Ages (and the cargoes of grain and cloth that aided the shipboard migration of rats and fleas) from the Black Sea, through the busy waters of the Mediterranean and out into the Atlantic, North Sea and Baltic, do much to explain plague's rapid dispersal and the main lines of its dissemination.

Similarly, plague, or rather the rats and fleas that bore the deadly bacillus, took advantage of the crowded and insanitary conditions common in European and Asian cities at the time. The spread of the black rat, and its fondness for living close to human dwellings, have often been cited as a factor favouring the repeated transference of plague from rodents to humans. It has equally been suggested that the changing nature of the urban environment, with greater use made of brick and stone rather

[11]   Ibid., pp. 141–84.

than wood, daub and thatch, helped separate people from rats and their pathogens and so diminished the frequency and severity of plague epidemics in seventeenth- and eighteenth-century Europe.

'Ripe for Catastrophe'

The interplay of human and environmental factors has also been used in a more instrumentalist fashion by those historians who have sought to relate the Black Death to underlying demographic and agrarian trends, and it is here perhaps that discussion of plague has been most readily incorporated into other, mainly Malthusian, agendas. It has been argued that rapid population growth had already brought Europe to the verge of an environmental and demographic crisis even before plague actually struck. As Le Roy Ladurie puts it, 'Everything in 1348' – the year the Black Death swept through France – 'was ripe for the unleashing of a catastrophe.'[12] Europe by the early fourteenth century had, it is said, over-extended itself: it was already a 'full world' and lacked the material means to support a population which had grown enormously over the previous three hundred years. In England, the population had trebled since the Doomsday survey of 1086, and by 1340 stood at an estimated four million. New land brought under the plough was of poorer and poorer quality, and the high yields at first gathered from newly cultivated land quickly dwindled in the absence of more effective farming techniques to maintain soil fertility. As a result, agricultural expansion was already grinding to a halt even before plague struck.

Climate may also have lent a hand. Colder, wetter conditions and resulting poor harvests helped push much of Europe beyond its fragile subsistence base and may have accounted for the famine that stalked Western Europe in 1315–16 as well as the severe

---

[12] Emmanuel Le Roy Ladurie, 'A concept: the unification of the globe by disease (fourteenth to seventeenth centuries)', in *The Mind and Method of the Historian*, Brighton, 1981, p. 51.

plague mortality thirty years later. 'What a magnificent field of action for the Black Death of 1348,' Le Roy Ladurie remarks, 'that holocaust of the undernourished'.[13] Philip Ziegler claims in similarly Malthusian mood:

> it cannot be denied that it [plague] found awaiting it in Europe a population singularly ill-equipped to resist. Distracted by wars, weakened by malnutrition, exhausted by his struggle to win a living from his inadequate portion of ever less fertile land, the medieval peasant was ready to succumb even before the blow had fallen.[14]

But the idea of the medieval peasant falling 'easy prey' to plague is too conveniently Malthusian. The disease was no respecter of class: it struck down the wealthy as well as the poor, the nobility and clergy as well as the peasantry; it attacked towns even more ferociously than the countryside. As already indicated, many peasants and labourers actually benefited from the labour shortage and enjoyed higher wages and improved living conditions as a result, and yet, despite these economic gains and a fall in the population to far below pre-plague levels, the heavy mortality continued. Plague can even be seen as 'a permanent debunking of the Malthusian model', for it was 'far from resolving the problems of a world which people have supposed to have been over-populated'.[15]

On the other hand, it has recently been argued that plague was 'largely exogenous, or external, to the socio-demographic system. It acted independently of modes of social organization, levels of development, density of settlement, and so on. The ability of plague to infect and kill bore no relation to one's state of health, age, or level of nutrition.'[16] This must surely be an over-statement. It would be hard to deny a connection between any

---

[13]   Emmanuel Le Roy Ladurie, *The Peasants of Languedoc*, Urbana, 1974, p. 13.

[14]   Philip Ziegler, *The Black Death*, London, 1982, p. 35.

[15]   Pierre Chaunu, *European Expansion in the Later Middle Ages*, Amsterdam, 1979, p. 62.

[16]   Massimo Livi-Bacci, *A Concise History of World Population*, Oxford, 1992, p. 48.

epidemic disease and patterns of social organization. More than the peasantry, it was ports, towns and cities that paid the heaviest price. They were often the initial point of epidemic invasion, and crowded and insanitary urban conditions enabled the disease to spread rapidly and to recur frequently. The movement of plague into the surrounding countryside often owed more to urban panic – with those who could afford to do so fleeing to the supposed safety of the countryside but in fact often taking the disease with them – than to rural destitution and malnutrition, or to the existence, as in Tuscany and East Anglia, of well-developed commercial networks and much-frequented trade routes. Some of the most prosperous port cities in Europe were repeatedly reinfected by plague and suffered heavily for their commerce with the East. As late as 1630–1 Venice, one of the wealthiest cities in Europe, is estimated to have lost a third of its population to plague, and it is surely no accident that Marseilles, a port with trading connections with the eastern Mediterranean, was the site of the last major outbreak in Western Europe.

The Black Death cannot be treated simply as an 'environmental accident', any more than it can be seen as just a Malthusian check on an over-populated Europe. Knowing how plague was caused and how the plague bacillus spreads from rats to humans, or from person to person, helps deepen our understanding of how the disease affected human societies, and why it affected different societies differently. But epidemiology alone cannot explain the various social and cultural reactions that plague produced or the ways in which the crisis affected long-term historical trends. Even if plague was in some respects a 'tragic accident at variance with the normal advance of events',[17] such an unpredictable event still had its social antecedents and evoked responses that were culturally predetermined. Plague was a profound shock but one, significantly, Europe was able to absorb and survive.

[17] Elizabeth Carpentier, quoted in Ziegler, *Black Death*, p. 34.

5

Crossing Biological Boundaries

Global Unification by Disease

The Black Death has assumed great, and increasing, importance
in the economic and social history of medieval Europe. But it has
also been taken as a model for the wider role of epidemic diseases.
The Black Death has been used to show how epidemics can have
a sudden, massive impact on the course of human history; but it
is also seen as a link in an expanding network of trans-oceanic
'exchanges', which involved people, plants and animals as well as
diseases.

It has been suggested by W. H. McNeill, Emmanuel Le Roy
Ladurie and others that the Black Death marked an important
stage in a process of 'microbial unification', the creation of a
single 'disease pool' shared between Europe, Asia and northern
Africa.[1] Smallpox, measles and plague are among a number of
diseases thought to have originated in Asia, where they had long
been resident among densely populated, urban-focused societies,
whose members lived in close proximity to domesticated animals
and to the parasites and vectors that shared their fields and

[1] William H. McNeill, *Plagues and Peoples*, 1st edn 1976, London, 1979;
Emmanuel Le Roy Ladurie, 'A concept: the unification of the globe by disease
(fourteenth to seventeenth centuries)', in *The Mind and Method of the Historian*,
Brighton, 1981.

dwellings. As long as the main centres of civilization in Europe and Asia remained effectively isolated from each other by jungles, deserts, mountains and seas, there was little likelihood of a disease that had evolved in one environment being successfully transferred to another. But military conquest, migration and long-distance trade, of the kind that brought India, China and Europe closer together during the early centuries of the Christian era, and again during the Mongol period, enabled such diseases as smallpox and plague to move out of their customary localities and establish themselves further west. Initially their impact was, in human terms, highly destructive, as the history of bubonic plague in Europe and the Middle East in the fourteenth century shows. But, over time, levels of human (or vector) resistance grew, or changes occurred in the environment, and such diseases became less formidable, subsiding from being agents of mass destruction to the status of childhood illnesses.

Following this confluence of Old World diseases, a process seen to have been largely completed with the Black Death, there followed a second stage which Le Roy Ladurie terms 'the unification of the globe by disease'. This occurred between about 1500 and 1700 with the beginning of oceanic exploration and trade through the European 'voyages of discovery'. Diseases by then well established in Eurasia slipped across the Atlantic to the New World, where they unleashed a wave of epidemic destruction comparable to that which Europe had recently experienced in the Black Death. In the view of some historians, the devastation was even greater, amounting to 'bacteriological genocide', as the Americas experienced in decades an exposure to new pathogens which Europe had known only piecemeal over the course of a thousand years of contact and contagion.

This New World episode was, in turn, the start of an even wider movement. In a further stage of 'global unification', beginning in the mid-eighteenth century, the same Old World diseases, especially smallpox, measles and influenza, made destructive inroads into other societies (Australia, New Zealand, the Pacific islands) brought within reach of European commerce and control. Meanwhile, new diseases arising out of Asia, like cholera in the early nineteenth century, also rapidly circled the globe, giving apparent confirmation to the fact that the world was by that stage truly united by its diseases.

Such trans-oceanic 'exchanges' and their human and biological consequences have attracted considerable attention in recent years. They have helped open up new approaches to world history and have gone a long way towards replacing once fashionable theories of climatic determinism. They appeal to a historical taste for the dramatic as well as the scientific, and have formed an important part of a growing literature in which environmental and biological factors are seen to have a profound, even determining, effect on human societies worldwide. Among the most influential writers in this genre have undoubtedly been W. H. McNeill, whose *Plagues and Peoples* was published in 1976, and Alfred W. Crosby, whose book *The Columbian Exchange* appeared four years earlier. A central theme in Crosby's account is the importance of biological factors in history. 'Man,' he declares, 'is a biological entity before he is a Roman Catholic or a capitalist or anything else'. 'The first step to understanding man is to consider him as a biological entity which has existed on this globe, affecting, and in turn affected by, his fellow organisms, for many thousands of years.' Crosby continues:

> Before the historian can judge wisely the political skills of human groups or the strength of their economies or the meaning of their literatures, he must first know how successful their member human beings were at staying alive and reproducing themselves. He must have some idea of how their efforts in accomplishing these tasks affected their environments as a whole and the other living things which were a part of these environments.[2]

Crosby illustrated this radical claim for the reorientation of history by highlighting the remarkable two-way traffic of plants, animals and diseases across the Atlantic following Columbus's arrival in the Americas in 1492. In addition to various animals

[2] Alfred W. Crosby, *The Columbian Exchange: Biological and Cultural Consequences of 1492*, Westport, 1972, pp. xiii-xiv.

and plants, from horses to wheat, from pigs to peaches, deliberately introduced into the Americas, Europe also unwittingly transferred to the New World such infectious diseases as smallpox, measles, typhus and influenza. Through the Atlantic slave trade, West Africa added to this deadly cargo of disease yellow fever and falciparum malaria. There was, however, not much of a return traffic in the epidemiological aspects of this 'exchange'. With the possible exception of syphilis, the Americas shipped few major pathogens back across the Atlantic to plague Africa and Eurasia. On the contrary, the New World contributed generously to the well-being of the Old, not only through the mineral bonanza of its gold and silver, but also by making available to the rest of the world such life-sustaining crops as potatoes, maize, beans and cassava, and a number of valuable medicinal drugs, including cinchona and ipecacuanha.

As far as is known, none of the major Old World diseases was previously present in the Americas because of its long isolation from Europe and Asia. Like childhood friends, the continents had long since drifted apart and lost all contact. The Amerindians' ancestors, immigrants from Asia via the Bering Strait, could only, Crosby reasons in Darwinian mode, have endured the cold crossing from Siberia to Alaska if they had been hardy and shed their most serious diseases: 'In the crudest sense, the life of the earliest Americans was a matter of the survival of the fittest.'[3] Thus, although they may have suffered from various nutritional and parasitic diseases, the native Americans are thought to have been largely free of the many infectious diseases that had stalked Eurasia for hundreds, even thousands, of years. The 'undeveloped level of Amerindian disease' is seen by Crosby and McNeill as part of the more general 'biological vulnerability' of the native Americans, a factor that had dire consequences for their capacity for survival once contact with Europe and Africa was opened up.

[3] Ibid., p. 31.

New World 'Holocaust'

The argument about the Amerindians' biological vulnerability is often linked to statements about the large populations thought to have inhabited the Americas on the eve of the Spanish conquest. Before the Second World War, it was assumed that the entire population of the Americas in 1492 was no more than eight to fifteen million. But, through the work of the demographers W. W. Borah and S. F. Cook, these estimates have been sharply revised upwards since the 1940s, until the Amerindian population on the eve of the conquest is put at between seventy-five and 100 million, with roughly twenty-five million each in the two most populous areas, Mexico and the Andes. From such high levels, demographic decline in the early decades of Spanish rule is seen to have been nothing less than catastrophic. By 1568, less than fifty years after the invasion of the Aztec empire by Hernan Cortes and his followers, the population of central Mexico had plummeted, Cook and Borah estimated, from twenty-five million to barely one million.

To an even greater extent than in Europe after the onslaught of the Black Death, demographic recovery was extremely slow among the native Americans – where they even continued to exist at all. In the West Indies, where again high population levels were thought to have existed before 1492, the fall was so catastrophic as to be irreversible. Of the seven or eight million Arawaks Cook and Borah estimated to have inhabited pre-conquest Hispaniola (the island today divided between Haiti and the Dominican Republic), by 1508 there were less than 100,000, and by the 1540s fewer than 500. This precipitous decline was matched, it is thought, by the experience of the other Caribbean islands and the tropical lowlands of Central America. By the middle of the eighteenth century, the population of native Americans had plummeted from the estimated seventy-five to 100 million of pre-conquest days to a mere 250,000. This has been called the 'worst human holocaust the world had ever witnessed'.[4]

[4]   David E. Stannard, *American Holocaust: Columbus and the Conquest of the New World*, New York, 1992, p. 146.

What caused this 'holocaust'? The very use of the term inevitably invites comparison with Hitler's genocidal slaughter of the Jews. For that sustained and deliberate act of mass destruction we acknowledge human responsibility: but was any one to blame for the Amerindian 'holocaust', or was it, tragically but unavoidably, another biological 'accident' like the Black Death?

For centuries, one of the explanations for the rapid decline of the native American population has been the 'black legend' of Spanish brutality and exploitation. In several respects the Black Death and the 'black legend' stand as rival interpretations of the two great 'holocausts' of the late medieval and early modern world, the one representing biological 'accident', the other human culpability. It is alleged that during the conquest and its immediate aftermath, the indigenous population was hunted down by the Spanish, plundered, deprived of food and forced to work in extremely arduous and life-sapping conditions on estates and mines. There are two main objections to this interpretation. One is that the 'black legend' was largely propaganda, produced by Spain's Protestant rivals or by critics like the Dominican friar and advocate of Indian rights, Bartolomé de las Casas. The other obstacle, more influential these days, is that the sheer size of the populations now envisaged for the Americas raises doubts about the capacity of Spanish cruelty alone to have caused such human carnage. In order to explain a demographic collapse of such haunting magnitude, some demographers and historians have brushed aside the 'black legend' and fixed instead upon the deadly potency of epidemic disease. McNeill concluded that

> human violence and disregard, however brutal, was not the major factor causing Amerindian populations to melt away as they did. After all, it was not in the interests of the Spaniards and other Europeans to allow potential taxpayers and the Indian work force to diminish. The main destructive role was certainly played by epidemic disease.[5]

---

[5]   McNeill, *Plagues and Peoples*, pp. 191–2.

There are several arguments supporting this epidemiological explanation. One is Malthusian. Despite their apparent success in adapting to the varied ecologies of the Americas and in domesticating such crops as maize and potatoes, the native Americans are said to have had a very limited subsistence base. According to McNeill, 'Amerindian populations were pressing hard against the limits set by available cultivable land in both Mexico and Peru when the Spaniards arrived'; or, as Le Roy Ladurie puts it, they had reached a point of 'Malthusian saturation'.[6] Central and northern South America on the eve of the conquest are seen to have been in much the same precarious position as the 'full world' of Europe on the eve of the Black Death, and the scale and precedent of the plague pandemic in Europe is often cited as evidence for the immense destructive capacity of disease in the Americas. Crosby, for instance, refers to an epidemic of smallpox, which he believes played a major part in the destruction of Amerindian populations in 1519–20, as having an 'influence on the history of America . . . as unquestionable and as spectacular as that of the Black Death on the history of the Old World'.[7]

The vulnerabilty of the Amerindians to epidemic assault is seen to manifest itself in other ways, too. The Aztecs and Incas had few domesticated animals, apart from guinea-pigs and llamas, and no flocks of sheep and cattle to slaughter as an emergency food larder. The general absence of domestic animals was a double disadvantage, for while the peoples of Eurasia had, over thousands of years, acquired many parasites and infections from their dogs, pigs and cattle, and learned to live with them, the Amerindians had not been subject to any such hardening exposure. Indeed, 'swine fever' may have the dubious distinction of being the first Old World epidemic to begin the depopulation of the Americas. An outbreak occurred in Hispaniola in 1493–4, caused by the importation of pigs from the Canary Islands on Columbus's second voyage to the West Indies.

Much is made by Crosby and others of the impact of such so-

6   Ibid., p. 188; Le Roy Ladurie, 'A concept', p. 72.
7   Crosby, *Columbian Exchange*, p. 42.

called 'virgin soil' epidemics – a phrase, whatever its scientific credentials, suggestive of the supposed biological innocence of native Americans in a predatory world. It is argued that a population with no previous experience of a disease, and hence no natural immunity to it, is particularly vulnerable and likely to suffer exceptionally high levels of mortality. By contrast, among a population long exposed to an infectious disease such as smallpox and measles, levels of mortality are likely to be low. People and microbes thus attain over time a degree of ecological equilibrium, of peaceful coexistence, which allows both to survive.

Crosby further argues that it is not only the nature of the disease itself which is important, but also how individuals react to it. The Aztecs were utterly unfamiliar with smallpox and other diseases that began to assail them with the Spanish invasion and had no idea how to treat them. They were psychologically, as well as biologically, disorientated. Indeed, like many other societies before and since, they barely recognized smallpox as a disease at all, but rather saw it as divine retribution. Lack of adequate nursing, the crippling effect of mass morbidity and mortality upon normal social and economic life, the inability even to harvest crops or draw water – all these were likely to intensify the impact of a 'virgin soil' epidemic.

Crosby argues that smallpox was particularly destructive, with a mortality in excess of 40 per cent among people experiencing it for the first time. It was a disease long established in Eurasia and those who encountered it there, often as a disease of childhood, acquired immunity not present among the Amerindians. Moreover, smallpox did not require fleas, insect vectors or any other intermediaries for its transmission. It spread through infected droplets of human breath, and so passed quickly from person to person in a way that could rapidly devastate a non-immune population. The consequence of this epidemiological onslaught, Crosby argues, was to overwhelm the Amerindians: political collapse and demoralization rapidly followed, clearing the way for European conquest and colonization. Smallpox thus provides him with a seemingly comprehensive answer to the question 'Why were the Europeans able to conquer America so easily?'

The rapid defeat of the Aztec and Inca empires by small bands of Spanish adventurers has long intrigued historians. How was Cortes, leading a few hundred men, able to seize the empire of Montezuma, who commanded several hundred thousand followers? The *conquistadores'* success has been attributed to many causes. One is the cultural and psychological susceptibility of the 'doom-laden' Aztecs and the opportunism of the invaders in exploiting divisions between the Aztec élite and its discontented subjects. There may be some justification to this argument. Other historians have seen it as a clash of technologies – between the swords, guns and armour of the Spanish and the 'Stone Age' weapons of their opponents – or the result of the fear and surprise caused by the Spaniards' horses and firearms. But of late the 'primitive' nature of Aztec and Inca society and its inability to cope with the small and ragged bands of invaders has come to be understood above all in biological terms. The successes of the *conquistadores* are attributed to the devastating effects of their 'biological allies', especially smallpox. Smallpox is thought to have spread from Spanish settlements on Hispaniola in 1518, entering Mexico with a second expedition that joined Cortes in 1520. At a critical juncture, when the Spanish had been driven out of the Aztec capital of Tenochtitlán, smallpox struck, killing thousands, preventing effective action by the defenders, and allowing Cortes to rally his forces and prepare for the final assault on the stricken city. 'Clearly,' McNeill observes, 'if smallpox had not come when it did, the Spanish victory could not have been achieved in Mexico.'[8] Epidemic smallpox is similarly seen to have undermined the resistance of the Incas of Peru, even before the arrival of the Spanish. The reigning Inca is thought to have died of the disease as did his son and heir, leaving no clear successor. Pizarro was able to exploit the ensuing disunity to achieve his own speedy conquest.

---

[8] McNeil, *Plagues and Peoples*, p. 192. For an interpretation of the Spanish conquest that effectively combines the biological with the cultural and psychological, see Tzvetan Todorov, *The Conquest of America: the Question of the Other*, New York, 1992.

Such bio-centric explanations, shared by many scholars who have written about the Spanish conquest and Amerindian depopulation, have tended to play down, even to deny, human culpability and conscious human agency. These massive 'die-offs' (as they are impersonally described) are seen as an ecological accident, not unlike the transference of plague from rats to people in fourteenth-century Eurasia. In an article that exemplifies this view, Donald Joralemon concluded that 'an uninvited third party' accompanied the Europeans when they first entered the Americas. 'No question of responsibility is appropriate here; no blame can be ascribed. The tragedy appears to be a necessary outcome of human interaction across biological boundaries.'[9] This 'holocaust', it seems, had no human author.

But doubts have been raised about the credibility of this biological explanation. Although few writers now question that disease played a significant role in both the Spanish conquest and the subsequent depopulation, some are more inclined to look for a broader combination of factors and to be wary of a single explanation for such a complex historical and demographic episode. Part of the problem is that the evidence for initial population levels and subsequent mortality in the Americas is far sketchier than for Europe at the time of the Black Death, where extensive written sources are available, and for this reason, among others, direct parallels with the Eurasian plague pandemic can be very misleading. It can reasonably be doubted, for instance, that the pre-conquest population of Mexico could ever have been anywhere close to the twenty-five million Cook and Borah claimed: five or at most, ten million has been suggested as a more realistic figure. If this were so, then neither the argument about extreme population pressure and Malthusian vulnerability before 1492, nor the supposed scale of the demographic collapse by 1600, would necessarily apply. Nor would it be quite so necessary to invoke disease, *deus ex machina*, to explain the sudden and extraordinary collapse of Amerindian societies.

[9]   Donald Joralemon, 'New World depopulation and the case of disease', *Journal of Anthropological Research*, 38, 1982, p. 125.

In arguing against Spanish brutality as a possible explanation for demographic collapse, Crosby remarks that European exploitation did not have time to destroy the Amerindians' health before smallpox and other diseases intervened. But Carl O. Sauer's judicious account of the early history of Hispaniola, first published in 1966, clearly showed how destructive of human life Spanish policies and practices were on the island from the very outset and how, as a result of warfare, massacres, forced labour, the destruction of indigenous agriculture, and the introduction of grazing and browsing animals, the population had fallen into headlong decline even before the arrival of smallpox in 1519. In the first decade of Spanish rule, Columbus's administration severely reduced the native population – 'in part by ineptness,' Sauer considered, 'in part by violent, ill-considered measures'.[10] The Spanish introduced a system of labour exploitation that could only be maintained 'by treating the natives as expendable'. It 'was not wanton brutality . . . that decimated the natives but a wrong and stupid system' of labour control and administration, the effects of which were exacerbated by the breakdown of the islanders' diet and social structure.[11] Within twenty-five years of Columbus's arrival in 1492, 'a well-structured and adjusted native society had become a formless proletariat in alien servitude, its customary habits and enjoyments lost.' Even the will to live and reproduce had gone.[12] In Hispaniola, and probably on the other West Indian islands as well, smallpox accelerated a process of rapid demoralization and decline, but did not initiate it.

As for Mexico, the main focus of Crosby's bio-centric discussion, Francis Brooks has looked in detail at the sources for the alleged smallpox epidemic of 1520, and is sceptical that they provide reliable historical evidence for such an outbreak. The main source used by later chroniclers and historians, Motolinia's *History of the Indians of New Spain*, owes much to the author's partisan spirit and an implicit analogy between the fate of the

[10]   Carl Ortwin Sauer, *The Early Spanish Main*, Berkeley, 1966, p. 155.
[11]   Ibid., p. 203.
[12]   Ibid., p. 204.

Aztecs and Biblical account of the plagues of Egypt. Brooks reckons that the real 'catastrophe' Tenochtitlán experienced in 1520–1 was war, not disease. Smallpox, if it occurred at all, or on any scale, was merely 'incidental'. While he can find no firm evidence for a massive smallpox epidemic killing a third or more of the Aztecs, there are data to support the view that thousands of people, perhaps as many as 100,000, died during the bitter fighting that accompanied the siege of the city in 1521.[13]

Doubts about an explanation of Amerindian collapse and depopulation based largely, if not exclusively, upon the impact of Old World diseases, have led to a shift away from environmental catastrophe to man-made disaster, or at least to a more balanced view of how disease and human behaviour might have operated together to bring wholesale death and destruction to native Americans. Even where it is accepted that disease may have been the immediate instrument of mortality, the underlying cause and the ultimate responsibility is seen to rest with human action – the rapacity of the Spanish invaders and colonists and their contempt for Amerindian life. Whereas Europe had had the time and opportunity to recover from the onslaught of the Black Death, and had not been confronted simultaneously by both human and microbial invasion, in the Americas the loss of land, the destruction of social systems and the effects of conquest and colonization prevented demographic recovery for centuries, by which time European power was firmly entrenched. The depopulation of the Americas was not therefore solely an ecological accident, the unintended consequence of 'virgin soil' epidemics and 'microbial invasion'. It was also the outcome of Europeans' racial contempt, brutal economic policies, and their lust for land and riches. Their ruthless pursuit of wealth amounted, in Stannard's damning phrase, to nothing less than 'purposeful genocide'.[14]

[13] Francis J. Brooks, 'Revising the conquest of Mexico: smallpox, sources, and populations', *Journal of Interdisciplinary History*, 24, 1993, pp. 1–29. For a response, see Robert McCaa, 'Spanish and Nahuatl views on smallpox and demographic catastrophe in Mexico', *Journal of Interdisciplinary History*, 25, 1995, pp. 397–431.

[14] Stannard, *American Holocaust*, p. xli.

'Ecological Imperialism'

In *The Columbian Exchange,* A. W. Crosby's main focus was on the Americas since 1492 and the two-way traffic in diseases, plants, animals and people between the Old World and the New. His *Ecological Imperialism,* published in 1986, represents a significant change of emphasis, as well as metaphor, from exchange to empire, and presents an even more expansive account of the biological dynamics behind European expansionism.[15] Crosby argues that the first attempts by Europeans to extend their influence beyond the shores of Europe were frustrated. The Norse adventurers and traders in the North Atlantic and North America were defeated by climate and by the absence of any 'biological weapons' to give them the edge over human and environmental adversaries. The Crusaders in the Levant were driven out not so much by military and political failures as by their inability to create a suitable ecological niche for themselves in an unfamiliar and, to their eyes, inhospitable landscape. Attempts by Europeans to establish themselves in the tropical regions of Africa, Asia and America were similarly doomed to failure: these, too, were lands 'within reach', but 'beyond grasp'. In terms of the diseases to which they were susceptible, the crops they tried to grow and the animals they attempted to raise, the Europeans had no biological advantages over the indigenes of Vinland, the Levant or the hot, moist tropics.

However, in the fourteenth and fifteenth centuries, when the Portuguese and Spanish turned their attention to the islands off north-west Africa, the situation proved strikingly different. In Madeira and the Canary Islands the impact of Old World biota – diseases, plants and animals combined – was dramatic and, once begun, irreversible. The islands were transformed by pigs, goats, rabbits and other animals which without predators, were free to devour the vegetation at will; by the introduction of wheat, vines,

---

[15] Alfred W. Crosby, *Ecological Imperialism: the Biological Expansion of Europe, 900–1900,* Cambridge, 1986.

sugar-cane and other crops; and by the arrival of European colonists to supplant the Guanches, the former inhabitants of the Canary Islands.

The ecological success of European colonization on the Atlantic islands was repeated on a far larger scale elsewhere, especially in those regions of the globe Crosby calls the 'Neo-Europes' – North America, Argentina and parts of Chile and southern Brazil, Australia and New Zealand – where native flora and fauna were to a great extent ousted by European plants and animals and where people of European descent have displaced the indigenes and now form the great majority of the population. Although thousands of miles from Europe itself, in these temperate lands Europeans, their crops and livestock, diseases, weeds and pests enjoyed the ecological advantage. Without predators to keep their numbers in check, pigs, sheep, cattle and horses flourished and reproduced themselves often at a far greater rate than in Europe. European grasses, cereal crops, vines and fruits proved prolific and underpinned the establishment of a European way of life. The 'Neo-Europes' were so successful agriculturally that in time they came to supply the world with a large proportion of its vital foodstuffs. Not all the biological introductions from Europe were intended or desired: weeds of European origin also set out on their own expansionary careers; pigs and rabbits devoured crops or gobbled up grasslands intended for human or livestock use. But, overall, the ecological balance had shifted decisively and profitably in the Europeans' favour.

The Struggle for Survival

As in *The Columbian Exchange*, so in his later book, Crosby identifies disease as the cutting edge of 'ecological imperialism'. He returns to the example of Cortes and the fall of Tenochtitlán, that most momentous episode in the history of European expansionism (though ironically Mexico did not join the ranks of the 'Neo-Europes'), but he also discusses the ravages of Old World diseases, especially smallpox, on the equally vulnerable populations of Australia and New Zealand. In Crosby's opinion, this epidemiological conquest explains how such small numbers of Europeans were able to triumph over vastly greater numbers of Aborigines and Maori as well as Amerindians, and how Europeans were able to transform the Americas and Australasia into 'Neo-Europes'.

The overwhelming message of *Ecological Imperialism*, as of *The Columbian Exchange*, is that Europe's relations with the rest of the world have been determined by factors over which Europeans had little practical or conscious control and hence no moral responsibility. The smallpox virus rather than Cortes was responsible for the destruction of the Aztecs; invading grasses, sheep and cattle colonized the plains of North and South America or New Zealand far more rapidly and effectively than the white settlers themselves. The 'success' of European imperialists overseas is accordingly ascribed to 'bio-geographical realities' rather than to the ambitions of those once hailed as conquerors. 'It was their germs,' Crosby writes, 'not these imperialists themselves, for all their brutality and callousness, that were chiefly responsible for sweeping aside the indigenes and opening the Neo-Europes to demographic takeover.'[16]

Crosby's biological determinism is clearly in many ways different from the earlier kinds of determinism we have considered. It is not principally about climate, or climatic change. Climate is not ascribed a directly determining effect, though, since the distinction between temperate and tropical lands is central to Crosby's whole argument, it does indirectly affect which areas of the globe are suitable for colonization by European biota. But significantly, in the 'Neo-Europes' it is not the pre-existing environment that does the dictating: the invading biota have the upper hand. It is not the continuities of the *longue durée* that count here or a long-evolving symbiosis between people and environment. Instead, it is the abrupt disjuncture that matters, the biological revolution that transforms the landscape and the human place within it. 'Ecological imperialism' rejects the old

[16] Ibid., p. 196.

determinism of Hippocrates and Montesquieu and replaces it with a newly aggressive, colonizing environmentalism.

There are, to be sure, a number of strengths to Crosby's argument. He effectively integrates the discussion of diseases with other factors of environmental change, notably plants and animals, to establish an ecological 'portmanteau', a whole set of importations that work together to transform the 'Neo-Europes' into biological replicas of Europe. Equally, Crosby's work helps to synthesize and popularize the work of a number of earlier geographers, botanists, ecologists and epidemiologists, and to bring their findings more firmly within the historical domain.

But, while respecting these achievements, Crosby's biological determinism must be open to criticism on several grounds. As with any essentially monocausal explanation, Crosby is tempted to attribute too much to a single set of forces and to ascribe too much to biology and too little to conscious human agency. As we have seen earlier in this chapter, it was not entirely a biological accident that the Spanish began the destruction of the native Americans, or that Europeans introduced their own familiar animals to lands they thought devoid of productive creatures. The intention to master, to Europeanize, was undoubtedly present, and was evident in the Americas from Columbus's second voyage in 1493–4 (and even in his first, two years earlier), though the outcome of such initiatives was frequently unpredictable. Crosby projects an air of biological inevitability about what was often the result of deliberate human intention.

There is also a powerfully Darwinian underpinning to *Ecological Imperialism*, as there often is when ideas of biological and human change are brought together in tandem. Crosby quotes, as one of the keynotes of his book, the remark Darwin made in 1836 when he visited Australia: 'Wherever the European has trod, death seems to pursue the aboriginal. We may look to the wide extent of the Americas, Polynesia, the Cape of Good Hope, and Australia, and we shall find the same result.'[17] We noted, too, in chapter 2 how Alfred Wallace, with Darwin co-founder of the

[17]   Charles Darwin, *Voyage of the Beagle*, Harmondsworth, 1989, p. 322.

theory of natural selection, not only elaborated this view with respect to competition between supposedly weaker and stronger races, but also made a connection, as Crosby does, between European colonization and the spread of European plants and animals at the expense of weaker native counterparts. Whether he is conscious of it or not, Crosby is heir to a tradition of scientific and historical scholarship that identifies European expansion with racial difference and the supposed inevitability of white racial supremacy.

One does not have to look far in the historical, geographical and epidemiological literature to find these sentiments expressed. They were spelt out, for instance, by P. M. Ashburn in the 1940s in a book cited by both Crosby and McNeill. Ashburn also argued that America was conquered 'largely by disease' and that the Amerindian was defeated 'not merely because he had no firearms, no horses, and no iron, but even more because he had no immunity to most of the diseases that the white man brought with him.' But Ashburn then significantly added that it was 'not fantastic to think that perhaps the white race has been and is great because of the difficulties it has mastered or partly mastered in the past.' Ashburn contrasted this success in the Americas with the slow advance and limited inroads of Europeans in Africa and Asia, where other races had the epidemiological advantages: the white man 'has dealt mainly with peoples who were his biological inferiors, in part at least, because of their lesser immunity to disease'.[18] Crosby may not use quite the same language, but in seeing the Amerindians, the Maori and the Australian Aborigines as 'biologically defenceless', he is advancing a similar explanation for their failure to survive competition from a 'superior' race.

The Darwinian view of nature in terms of the survival of the fittest is one that Crosby seems fully to endorse – in human as well as biological terms. One consequence of this is that he spares no retrospective glance for the great loss of human lives and the rich and diverse societies squandered by the advance of 'ecological

[18]   P. M. Ashburn, *The Ranks of Death: a Medical History of the Conquest of America*, New York, 1947, pp. 210–12.

imperialism'. He ignores the remarkable success of pre-Columbian societies in overcoming their own environmental challenges
– in domesticating maize and potatoes, crops of which Europe,
Asia and Africa were ultimately the beneficiaries – or in creating
elaborate terraces and water channels to make productive the
steep Andean hillsides. Crosby marches with the armies of conquest; he does not tarry with nature's losers. 'Ecological imperialism' seems to be less about loss and more about triumph.

One can also look beyond the immediate geographical and
historical limits Crosby sets himself to doubt the validity of many
of his ecological propositions. Was it only in the temperate 'Neo-
Europes' that Europe found 'success'? In India, in Africa, and in
many other parts of the world, Europeans encountered conditions
even more unfamiliar or environmentally unfriendly to them than
the Crusaders' Levant, and yet they were undeterred. Instead,
they developed other ways of gaining mastery of the land and its
products than the introduction of white settlers or the dispersal of
European plants and animals. They set up plantations to grow
crops Europe itself could not supply; they transferred plants from
one part of the tropics to another to facilitate their commercial
exploitation. In areas where Europeans did not thrive, or chose
not to labour, they mobilized slaves and indentured labourers to
work for them; they recruited sepoy armies and native levies to
fight on their behalf and thus shoulder the double burden of
warfare and disease. In places outside the 'Neo-Europes' where
they did work and settle, Europeans devised medical and sanitary
means to protect them from the worst effects of an alien environment. And in areas where disease and the whole awesome
armoury of European biota failed to do the trick for Europe
quickly or completely enough, recourse to calculated genocide
might follow, as the fate of the last Tasmanians in the 1830s
illustrates. 'Ecological imperialism' seems in the final analysis to
be a strangely one-dimensional, Eurocentric reading of the processes of European expansionism, a cautionary tale in the dangers
of ascribing to biology what more properly belongs to human
agency.

Slavery

The Black Death in Europe and the depopulation of the Americas have been seen as analogous phenomena, the role of plague in reducing the population of Europe by a third or more paralleled by the decimation, in some cases the extinction, of the Amerindians. They are also seen to be linked phenomena. The Black Death reputedly marked the completion of the unification of the disease pools of Eurasia; the depopulation of the Americas – by smallpox, measles, typhus, influenza and other diseases – marked the extension of this unifying process beyond the Atlantic, to the Americas, and subsequently to Australasia and the Pacific. And there is a further link between the two – slavery.

The argument is twofold. First, until the Black Death, Europe had supplemented its population with slaves from Asia and the Black Sea region in particular. The plague pandemic interrupted this source of supply while (contrary to the general tenor of the Malthusian argument about an over-populated Europe) increasing the demand for slaves to replace the lost populations, especially in Portugal. Europe began to turn to West Africa as a source of captive manpower to meet its needs and, by the fifteenth and sixteenth centuries, those of the expanding economies of the Atlantic islands and the Americas. In this way two momentous human and ecological events – the Black Death and the advent of the African slave trade – became historically conjoined.

A second line of argument takes up the story on the other side of the Atlantic. The rapid decline of the Amerindian populations of the West Indies and mainland Central and South America left the Spanish the other Europeans facing a major shortage of labour. In order to unlock the wealth of the New World they needed huge armies of labourers – to work in the deadly silver mines of Potosi and Zacatecas and to meet the seemingly insatiable demand for plantation and estate labour, especially with the rapid development of sugar as a plantation crop in the mid-seventeenth century. Although at first Europeans were present as labourers and small farmers on a number of Caribbean islands, with the advent of the sugar revolution West Indian society became increasingly polarized between black slaves and a white planter class. In Barbados, the leading sugar producer in the West Indies at the time, the number of whites declined from 30,000 in 1650 to 15,500 fifty years later, while the number of African slaves rose to more than 41,000. The Atlantic slave trade, which had begun as early as 1518, had thus by the late seventeenth century grown to massive proportions and was to continue, despite the abolition of Britain's share of the slave trade in 1807, until the second half of the nineteenth century. It is estimated to have resulted in the enslavement and transportation of some twenty million Africans.

That Africa was – and for centuries remained – the source of this vast and inhuman traffic has been explained in several ways: the geographical proximity of West Africa to the slave-seeking societies of the Caribbean, Brazil and the southern United States; the fact that Africans, versed in the arts of settled agriculture and hoe cultivation, had the skills needed for plantation work; the prior existence of forms of slavery in West Africa that the European traders could exploit and nurture through the sale of firearms, cloth and other trade goods; and more simply the racism of Europeans, who regarded 'blacks' and 'pagans' as barely human and their lives as being of little more than commercial value.

Historians continue to debate the importance of these and other explanations. But an important adjunct to these theories has grown out of the burgeoning interest in the history of disease and its perceived relevance to European expansionism. It has been argued, for instance, that although the Portuguese and other Europeans had commercial contact with the West African coast from the mid-fifteenth century onwards, they were prevented as much by disease (yellow fever and malaria) as by political and military resistance from extending their influence inland. Even the slave trade took a heavy toll of the European traders, soldiers and sailors who visited the coast or temporarily resided in forts along the West African shoreline. Not until the nineteenth century, it has been claimed, with the increasing use of quinine to keep malaria at bay, were Europeans able to establish a more permanent presence in West Africa. The extraction of African labour via the slave trade for service in the New World was thus one of the few means by which Europe could tap the resources – especially the human resources – of the region.

But it is also argued that Africans had certain biological assets which made them better able to survive in the West Indies and on the plantations of mainland America than either the Amerindians or immigrant whites. Why did the Caribbean and neighbouring parts of tropical America not become, in Crosby's terms, a 'Neo-Europe', despite the initial impact of Old World diseases, plants and animals? The first generation of European conquerors and settlers found the region relatively healthy and were little affected by Old World infectious diseases, against which the Amerindians were 'biologically defenceless', but in time the disease environment turned decisively against them. Recurrent epidemics of smallpox exacted a heavy toll among settlers who had had no childhood exposure to the disease, and so had no acquired immunity. Even more significantly, the contacts opened up via the slave trade between West Africa and the Americas introduced to the Caribbean diseases that were equally deadly for Europeans. Pre-eminent among these was yellow fever, a mosquito-borne disease long established in West Africa, but not generally found north of the Sahara, and malaria of a far more deadly strain (falciparum malaria) than the vivax malaria, or ague, with which early modern Europe was familiar.

By the end of the seventeenth century, as large numbers of African slaves began to be imported into the region, yellow fever began to affect the sugar islands of Barbados and Guadeloupe, before moving on to Cuba and the Mexican coast. By the mid-eighteenth century yellow fever was causing high levels of mortality among the white population of the West Indies, especially among those, like newly arrived soldiers and sailors, who had not been 'seasoned' by earlier attacks. Outbreaks of yellow fever seriously harassed the naval and military campaigns fought between the English, Spanish and French during the middle and later decades of the eighteenth century. Between 1793 and 1796 the British lost 80,000 troops in the West Indies (more than in the whole of the Duke of Wellington's Peninsular War campaigns), more than half of them to yellow fever. When the French sent an army under Leclerc, Napoleon's brother-in-law, to Santo Domingo (Hispaniola) in 1802 to retake the island, they lost 40,000 officers and men to yellow fever in ten months, including their commander.[19] The subsequent withdrawal of French forces allowed the successful establishment of the first independent black nation in the Americas in Haiti. In the face of such unacceptable losses, Napoleon abandoned his ambitions in the New World altogether, selling Louisiana to the United States in 1803. The role of yellow fever in the politics of the region has thus been seen as 'determining the outcome' of military operations in the region and the wider pattern of European colonization, a further illustration of the power of epidemic disease to shape the course of human history.[20] (It should be noted, though, in passing that this severe mortality among whites did not go unanswered: it was partly met by raising black regiments to fight in the West Indies in the 1790s, and partly by putting white troops in barracks built in more salubrious locations.)

But while European sickness and mortality from yellow fever and malaria soared, slaves of African origin were found to be largely immune to the diseases. This African resistance to disease, K. F. Kiple has observed,

> made blacks even more valuable as slaves. The Indians died of European and African diseases, the whites died of African diseases, but the blacks were able to survive both, and it did not take the Europeans long to conclude that Africans were especially designed for hard work in hot places.[21]

[19] Kenneth F. Kiple and Kriemhild Conee Ornelas, 'Race, war and tropical medicine in the eighteenth-century Caribbean', in David Arnold (ed.), *Warm Climates and Western Medicine: the Emergence of Tropical Medicine, 1500-1900*, Amsterdam, 1996, pp. 70–1.
[20] Francisco Guerra, 'The influence of disease on race, logistics and colonization in the Antilles', *Journal of Tropical Medicine and Hygiene*, 69, 1966, pp. 23–35.
[21] Kenneth F. Kiple, 'Disease ecologies of the Caribbean', in K. F. Kiple (ed.), *The Cambridge World History of Human Disease*, Cambridge, 1993, p. 499.

In the tropical lowlands of the New World malaria and yellow fever not only accelerated Amerindian mortality, but also prevented sustainable white populations. African slavery thus replaced free white labour wherever the impact of disease was most critical. But one must not forget that with the advent of large sugar estates white labour was already in decline in the Caribbean. It was the mass importation of slaves to meet the insatiable demand for plantation labour that caused the adverse epidemiological conditions, not the other way round. It originated in a cultural choice founded on the economics of sugar production and the relative value placed on white as against black lives and freedom. Sugar was not native to the region either, but was a crop which had its immediate origins in Mediterranean Europe and the Atlantic islands.

Nor should it be imagined that Africans somehow escaped the heavy burden of mortality its production entailed. African lives were lost in prodigious numbers in the initial act of enslavement, in coastal depots in West Africa, in crowded and disease-ridden slave ships, and in the Americas themselves from diseases such as yaws, leprosy and filariasis, imported from Africa. Even smallpox may increasingly have come from West Africa rather than from Europe, and 'the Caribbean was transformed practically overnight into an extension of an African, as opposed to a European, disease environment.' The slaves also suffered from appallingly low levels of nutrition, 'probably the biggest destroyer of slave life' on the plantations.[22]

None the less, it is argued that in many ways Africans were more 'suitable' than whites to live and work in the West Indies. Biology, not white racism or the economics of sugar production, is thus seen to provide the fundamental explanation for the centuries-long importation of black slaves. But, as in the earlier discussion of Crosby's work, it is difficult not to feel uncomfortable about neo-Darwinian arguments that stress the biological 'weakness' of one race (the Indians) bound for extinction, the 'fitness' of another (the Africans) for slavery, and that of a third

[22] Ibid., pp. 500–1.

(the Europeans) for mastery and exploitation. It smacks too much of biological determinism, a retrospective rationalization for the enslavement of millions of Africans, and a convenient freeing of Europeans and white Americans from the burden of moral responsibility for slavery.

6
The Ecological Frontier

The Frontier Thesis

It is possible to think of environmental history as being about two relatively fixed and largely independent elements – the human and the natural – working and interacting together over a great expanse of time. But what happens when this symbiotic relationship breaks down or is fundamentally altered because the environment itself changes or because people migrate to new, environmentally very different lands? Although much environmental writing has focused upon the contrast between fixed societies habituated to different climates and topographies (as classically in Montesquieu's contrasts between Europe and Asia), there has always been a concern with what happens to the natives of one locality when they move to another (as in the discussion in the first part of *Airs, Waters, Places*). In that text it was assumed that individuals moving from one location to another became subject to new environmental forces – unfamiliar winds, strange waters – and that their bodies were affected (or afflicted) accordingly.

Conversely, theories of migration and colonization have often stressed the radical transformation of the environment brought about by the arrival of outsiders. In an extreme version of this, Crosby's *Ecological Imperialism*[1] argues that, in the temperate parts

---

[1] Alfred W. Crosby, *Ecological Imperialism: the Biological Expansion of Europe, 900–1900*, Cambridge, 1986.

of the Americas and Australasia, the arrival of the Europeans resulted in substantial changes to the environment as Old World plants, animals and diseases embarked on their own process of colonization, thus making it possible for Europeans to establish themselves successfully and maintain a largely European way of life. Environments, then, are not necessarily fixed entities to which outsiders have either to accommodate themselves or perish. They are not, as it were, rooted to one spot: they, too, can cross the seas, march with armies, and conquer entire continents.

We are presented, then, with two alternative possibilities. Are immigrant societies forced to adapt and conform to a local environment, or do they bring their own environmental baggage with them? Does the 'challenge' (to echo Toynbee) of a new environment force the abandonment of old cultural ideas and practices and stimulate the creation of those more appropriate to the new? Or are cultural traditions so firmly entrenched that they are hard to shed and survive even in seemingly hostile territory?

One of the ways in which historians have addressed the issue of fixity and change, of old culture encountering new environment, has been through the 'frontier thesis' as initially formulated by the American historian Frederick Jackson Turner. Although much of the discussion of the Turner thesis has been built around debates within North American history, it serves, like the discussion of the Black Death or disease and depopulation in the Americas, as a model that can be applied or discussed in relation to a number of other 'frontier societies' as well. One can argue more generally that the idea of an advancing frontier is one of the principal ways in which historians have sought to conceptualize the process of interaction and conflict between two culturally distinct sets of people and the environmental ideas and practices they represent. The frontier has, moreover, served as a device, partly derived from Darwinian ideas of evolution, by which to express the dynamic character of European (or Europe-derived) societies and to contrast them with the supposedly primitive and unchanging societies and ecologies they displaced.

Over a century ago, in July 1893, Frederick Jackson Turner presented his celebrated paper on 'The Significance of the Frontier in American History' to the American Historical Association. The thesis Turner presented has aptly been called 'an environmental explanation of Americanism';[2] it has certainly proved one of the most influential statements of environmental determinism in human history. Turner began by noting a recent bulletin issued by the Superintendent of the Census which, in effect, declared the end of unsettled land in the United States and thus, in Turner's view, the end of the American frontier. This represented

> the closing of a great historic moment. Up to our own day American history has been in a degree the history of the colonization of the Great West. The existence of an area of free land, its continuous recession, and the advance of American settlement westward, explain American development.[3]

Turner was reacting against the views of those historians, including his own former teacher Herbert Baxter Adams, who believed the origins of America's institutions, including its democracy, lay in Europe and especially, as was common at the time, in early Germanic folk institutions. Turner was at pains to reject this connection and assert instead the uniqueness of the American historical experience. In most nations, he claimed, the development of social institutions had occurred within a fixed geographical area, or, if it expanded, the nation had met with, and had to conquer, other peoples. By contrast, the United States had evolved out of a situation in which 'free' land was always available, and the frontier, 'the meeting point between savagery and civilization', had constantly moved further and further west as older territories were brought under a more settled system of agriculture and government. Out of this interaction with the frontier environment came a constant process of rebirth and renewal, as Americans shed inherited ideas and institutions and adapted to the challenge and stimulus of a new situation. Hence, Turner

[2] David M. Potter, *People of Plenty: Economic Abundance and the American Character*, Chicago, 1954, p. 22.
[3] 'The significance of the frontier in American history', in Frederick Jackson Turner, *The Frontier in American History*, New York, 1953, p. 1.

declared, 'the history of this nation is not the Atlantic coast' – the area of first settlement and the site of geographical and cultural links with the Old World – 'it is the Great West', and in his view all the major issues of American history, slavery included, could not be adequately understood unless this fundamental fact were duly recognized.

Turner has been widely criticized for the looseness with which he used the term 'frontier'. Despite its centrality in his thought, he used it, one recent historian remarks, 'carelessly, or rather indiscriminately'.[4] Did he mean a particular geographical location – the western two-thirds of the United States as opposed to the eastern seaboard? Or did he mean process rather than place, a stage of historical evolution through which every part of what became the United States was eventually to pass before moving on to a more settled way of life? Turner's elusiveness on this point has both inspired and infuriated subsequent generations of historians. But from our perspective it is important to see the frontier as signifying an amalgamation of physical and cultural forces, as Turner himself indicated in 1893:

> The frontier is the line of most rapid and effective Americanization. The wilderness masters the colonist. It finds him a European in dress, industries, tools, modes of travel, and thought. It takes him from the railroad car and puts him in the birch canoe. It strips off the garments of civilization and arrays him in the hunting shirt and the moccasin. It puts him in the log cabin of the Cherokee and Iroquois and runs an Indian palisade around him. Before long he has gone to planting Indian corn and ploughing with a sharp stick; he shouts the war cry and takes the scalp in orthodox Indian fashion. In short, at the frontier the environment is at first too strong for the man. He must accept the conditions which it furnishes, or perish . . . .[5]

But Turner is not recalling Rousseau. There are no noble savages here. The frontier forces the immigrant to shed his

[4]   Roger L. Nichols (ed.), *American Frontier and Western Issues: a Historiographical Review*, New York, 1986, Introduction, p. 2.
[5]   Turner, 'The significance of the frontier', pp. 3–4.

European past (much as Francis Parkman described migrants on the Oregon Trail jettisoning 'ancient claw-footed tables' and 'massive bureaus of carved oak' as they headed west in wagon trains across the prairies).[6] The burden of inherited customs and conventions is cast aside, and for the moment 'the environment is . . . too strong for the man.' But he does not for ever remain at the level of Indian savagery, taking scalps and paddling birchbark canoes. Gradually, the frontiersman (and Turner thought mostly of men, not women, being exposed to, and transformed by, the frontier situation) begins to reverse the balance between environment and culture and to 'transform the wilderness'. Having learnt what he needed to learn from the Indian for his own survival, he soon outstrips and supplants him. Beginning on the East Coast, 'still the frontier of Europe in a very real sense', the line of advance moved further and further west, becoming, in the process, 'more and more American'. 'Thus the advance of the frontier,' Turner wrote, 'has meant a steady movement away from the influence of Europe, a steady growth of independence on American lines.' Even when the frontier moved on, and left settled regions, with farms, roads and towns, behind in its wake, those regions retained something of the character of the earlier frontier experience.

The successive stages in the westward advance of the frontier bore the imprint of certain 'natural boundary lines', which 'served to mark and to effect the characteristics' of the moving frontier. In the seventeenth century, the frontier stood at the 'fall line' on the eastern seaboard; in the eighteenth century, it had moved as far west as the Allegheny Mountains. By the first quarter of the nineteenth century the frontier had advanced to the Mississippi and (the leap forward to California apart) stood by mid-century at the Missouri, finally reaching the arid grasslands and the Rocky Mountains in the 1880s and 1890s. These were not only natural hurdles to be crossed and physical obstacles to be overcome; they also marked successive stages in the long struggle against the

[6] Francis Parkman, *The Oregon Trail: Sketches of Prairie and Rocky-Mountain Life*, 1st published 1849, London, 1944, p. 60.

Indians. The Indian wars provided for Turner the stimulus of a unifying force: they brought the states together in united action; they acted as a military training school; and they served to develop 'the stalwart and rugged qualities of the frontiersman'. In Turner's view the Indians were part of the environment, kin to nature, as primitive and obstructive, but also as formative a force as the forests, rivers and mountain passes. They had to be overcome, thus enabling white America's institutions and values to flourish in their stead.

The Evolving Frontier

Turner was arguing, on the one hand, against the racial and geneticist arguments of the Adams School, and their emphasis upon the European origins of American institutions ('Germanic germs' he called them disparagingly); but he was, on the other hand, endorsing the idea of Indians as an inferior people who needed to be conquered and excluded from the subsequent history of white America. The frontier was a frontier of racial subordination and then exclusion. In this there was a powerful and consciously Darwinian thrust. The frontier was not only the cultural matrix from which the new America was born; it was also a 'record of social evolution,' as Turner called it, from primitive Indian to civilized white factory, farm and courthouse. In a celebrated passage in his 1893 paper, which clearly identified this evolutionary process at work, Turner remarked:

> Stand at the Cumberland Gap and watch the progress of civilization, marching single file – the buffalo following the trail to the salt springs, the Indian, the fur-trader and hunter, the cattle-raiser, the pioneer farmer – and the frontier has passed by. Stand at South Pass in the Rockies a century later and see the same procession with wider intervals in between.[7]

---

[7]   Turner, 'The significance of the frontier', p. 12.

The part which the major river systems had played in opening up the West and helping to determine the location of its major ports and cities, and the way in which old Indian trails had become the commercial highways of the new America, caused Turner to reflect that 'in this progress from savage conditions lie topics for the evolutionist'. And three years later, in an essay on 'The Problem of the West', in language which reflects the imprint of evolutionary and ecological ideas on historical thinking by the 1890s, Turner remarked that the history of America's political institutions and democracy was not a history of imitation and borrowing from Europe, but 'a history of the evolution and adaptation of organs in response to changed environment, a history of the origin of new political species'.[8]

From the formative frontier experience was born the spirit of American democracy, epitomized for Turner by Andrew Jackson of Tennessee, whose democracy 'came from no theorist's dreams of the German forest', but issued 'stark and strong and full of life, from the American forest'. From the frontier territories came, too, the democratic franchise of the early nineteenth century and the egalitarianism of Western ideas and institutions compared to those of the more aristocratic, Europe-looking East. A sense of national identity was forged there too. It was through the 'crucible of the frontier' that immigrants from various European lands were brought together and merged into a 'composite nationality', there that they were 'Americanized, liberated, and fused into a mixed race, English in neither nationality nor characteristics'. Hence, 'the growth of nationalism', like the evolution of American political institutions, depended upon the advancing frontier.

But, correspondingly, the passing of the frontier in the 1890s, was for Turner a matter of regret and concern. If the frontier environment had played such a critical part in creating the American spirit and American institutions, its closing must threaten their future vitality, unless the tide of American power

---

[8] 'The problem of the West', in Turner, *The Frontier in American History*, p. 206.

were destined – as might yet be the case with the expansionism of the 1890s – to sweep yet further afield. In his 1896 essay on 'The Problem of the West', written only months before the outbreak of the Spanish–American War turned the attention of the United States to the Caribbean and Pacific, Turner concluded:

> For nearly three centuries the dominant fact in American life has been expansion. With the settlement of the Pacific coast and the occupation of the free lands, this movement has come to a check. That these energies of expansion will no longer operate would be a rash prediction, and the demands for a vigorous foreign policy, for an interoceanic canal, for a revival of our power upon the seas, and for the extension of American influence to outlying islands and adjoining countries, are indications that the movement will continue.[9]

Written shortly before two other Americans, Ellen Semple and Ellsworth Huntington, produced their versions of geographical determinism, Turner was articulating the spirit of a race-proud, imperially minded age. His thesis was formulated in the context of 'a nationalist ethos permeated by social Darwinism and the rise of the United States to the status of a major world power'.[10] But in looking back on the frontier and the 'free land' it had for centuries provided, Turner was also giving expression to the idea that North America had already become a 'closed space' and the apprehension that the old frontier spirit could not be rekindled in the rain-forests and savannahs of Central America or island south-east Asia. He surely sensed, as Semple and Huntington did, the looming environmental as well as political constraints on American expansion into culturally and physically alien territory. The expanding frontier of the north was running up against its tropical antithesis; the otherness of the 'tropicality' we will return to in chapter 8. After three hundred

---

[9] Ibid., p. 219.
[10] Howard Lamar and Leonard Thompson, 'Comparative frontier history', in H. Lamar and L. Thompson (eds), *The Frontier in History: North America and Southern Africa Compared*, New Haven, 1981, p. 4.

years of moving west across temperate America, there were now, it seemed, no 'new lands' left to occupy – or, at least, worthy of the conquest.

Expanding the Frontier Thesis

Turner's 'frontier thesis' has had a remarkable impact on the writing and conceptualization of American history, endowing it with a strong sense of an unfolding chronology and national cohesion. From the turn of the century until the early 1930s the Turner thesis was, not surprisingly, 'virtually unquestioned as the Holy Writ of American historiography'.[11] Thereafter, perhaps inevitably, a reaction set in and Turner was criticized, especially by left-looking historians for the many factors he had left out, such as the role of class conflict, urbanization and industrialization in shaping America's historical destiny. Writing in 1949, Richard Hofstadter fiercely attacked the 'frontier myth' for embodying the 'predominant American view of the American past', and for failing to consider the many other factors that had influenced the making of America's past. He was one of many historians who did not accept a backwoods origin for the democratic ideals of the American Revolution. In his view, they did not arise from an encounter with the wilderness but 'derived from the English republicanism of the seventeenth century – notably, of course, from Locke'. It was the cities that received the bulk of the immigrants, not the countryside, and it was there, in the urban melting pot, not the frontier 'crucible', that the process of building a nation truly began.[12]

By the 1960s Turner's reputation as a historian had sunk to a low ebb. His representation of the West and its part in America's history was deemed inaccurate and irrelevant, or decried – where it was taken seriously at all – as 'racist, sexist and imperialist' in

[11] Ray Allen Billington (ed.), *The Frontier Thesis: Valid Interpretation of American History?*, New York, 1966, Introduction, p. 3.
[12] Richard Hofstadter, 'Turner and the frontier myth', in ibid., pp. 100–6.

its depiction of western expansion and settlement.[13] Since then, however, there has been a marked, if qualified, revival of interest, partly because of the environmentalist thrust seen to lie behind the Turner thesis. Even if the frontier did not make America in quite the sense Turner envisaged, it was clearly the stage for many of its formative episodes and expressed many of its highest ideals and wildest delusions. One historian has gone so far as to claim that environmental history in the United States is in many ways 'a modern extension of the logic of Turnerian frontier historiography', though, as the same writer also points out, Turner's celebration of the destruction of the wilderness and its wildlife, and the conversion of 'free' land to agricultural and industrial use, appears from a present-day perspective to be the height of environmental bad taste.[14] Turner's history can now be seen as an account of a process of remorseless and irreversible environmental change with few parallels for its sheer rapidity and scale in the whole of history.

Although Turner stressed the uniqueness of the American experience and made few comparative observations even within the Americas (his blindness to the Canadian and Spanish American experience is one of the grounds on which he has been most frequently criticized), his thesis has been used as a model for the investigation and analysis of other regions undergoing broadly comparable processes of cultural and environmental change associated with European expansion since the fifteenth century – such as Canada, Latin America, South Africa and Australia – though never with quite the impact and assurance the Turner thesis has enjoyed at home. Part of the appeal of the thesis within America itself – for both disciples and detractors – has been the broad prospectus it provides for the discussion of the history of the United States as a whole, while at the same time providing – in

[13]   William Cronon, George Miles and Jay Gatlin, 'Becoming West: towards a new meaning for Western history', in W. Cronon, G. Miles and J. Gatlin (eds), *Under an Open Sky: Rethinking America's Western Past*, New York, 1992, p. 4.
[14]   John Opie, 'The environment and the frontier', in Nichols (ed.), *American Frontier*, p. 7.

the moving frontier – a yardstick by which to measure local processes of change.

Although Turner personally eschewed comparative history, it might be argued that the nature of his hypothesis and the boldness with which he presented it have had a profound effect on subsequent generations of historians seeking to locate the history of North America within world history, and their concern to do so not simply through a history of ideas, institutions and migrations, but also through a history of ecological context and impact. For Turner, of course, this was a history of America's uniqueness: the forests and frontier made the United States distinct from Europe and presumably from other societies even in the Americas. But other historians since Turner's day have rejected his ecological chauvinism while retaining the broad appeal of environmental determinism. In Crosby's *Ecological Imperialism*, for instance (see chapter 5), the frontier, still the seminal experience from which the new America arose, becomes an ecological frontier, an advancing tide not merely of human pioneers but also of Old World plants, animals and above all pathogens, sweeping away the indigenes and establishing their settlements not just across the Alleghenies and into the Great Plains but throughout the 'Neo-Europes'. America and the other regions of the globe with which Crosby is concerned represent, as did the West for Turner, 'virgin lands', societies culturally or biologically less advanced than the Old World, and therefore open to effective invasion. Turner's America and Crosby's 'Neo-Europes' proceed from a shared search for historical and ecological exceptionalism. They draw a potent contrast between what happens in 'virgin lands' and what happens either in Eurasia or in the tropics, where other cultural and environmental imperatives apply. It is thus on the basis of the North American experience that they seek to define the wider patterns of history, whether to make generalities or to plead for exceptions.

Europe's 'Great Frontier'

American historians have sought other ways to expand or amend the Turner thesis. One of the most ambitious of these was Walter Prescott Webb's *The Great Frontier*, first published in 1952. Webb paid generous tribute to the importance of Turner's work for American historiography, claiming that if every nation had but one contribution to make to world scholarship then in his view America's would be the frontier thesis. He added, however, that the concept needed to be lifted out of its narrowly North American setting and applied to the evolution of Western civilization as a whole since 1500. But while expanding the geographical parameters of the frontier idea to dimensions not dissimilar to Crosby's 'Neo-Europes', Webb retained Turner's sense of the value of 'free' land and the identification of indigenes with primitiveness and nature. He wrote:

> The concept of a moving frontier is applicable where a civilized people are advancing into a wilderness, an unsettled area, or one sparsely populated by primitive people. It was the sort of land into which the Boers moved in South Africa, the English in Australia, and the Americans and Canadians in their progress westward across North America. The frontier movement is an invasion of a land assumed to be vacant as distinguished from an invasion of an occupied or civilized country, an advance against nature rather than against men ... Inherent in the American concept of a moving frontier is the idea of a body of free land which can be had for the taking.[15]

But where Turner turned his back on Europe and Germanic folk origins, Webb (perhaps mindful of the debt Europe owed to the military and economic resources of the United States in the Second World War and its aftermath) saw Europe and the frontier societies as sharing a common destiny. They were the joint beneficiaries of the frontier experience. Columbus's discovery of

---

[15] Walter Prescott Webb, *The Great Frontier*, 1st edn 1952, Austin, 1964, p. 3.

the Americas thus becomes, by his account, a critical moment as much in the history of Europe as of the New World (a view European historians have tended to resist). For Webb, the frontier was 'almost if not quite as important in determining the life and institutions of modern Europe as it was in shaping the course of American history'. Without the frontier, modern Europe would be 'so different from what it is that it could hardly be considered modern at all'.[16]

His starting point was the kind of bleak Malthusian prospect of late medieval Europe we encountered in chapter 4, and which surely found its echoes in the war-ravaged Europe of the late 1940s and early 1950s when Webb was formulating his version of the frontier thesis. In 1500 Europe was a society of 100 million people, relatively stable but hard-pressed for the means of subsistence: it was poor and hungry, and had barely emerged from the trauma of the Black Death. Then, with the discovery of the Americas, 'came the miracle that was to change everything, the emancipator bearing rich gifts of land and more land, of gold and silver, and of new foods for every empty belly and new clothing stuffs for every half-naked back'. The riches of the Great Frontier, beginning with the Americas, but continuing with South Africa, Australia and New Zealand, were enough, Webb argued, 'to make the whole Metropolis rich'.[17]

In terms of three critical factors – population, land and capital – the Great Frontier shifted the balance decisively in Europe's favour. In 1500 there were twenty-seven people for every square mile of European territory. The opening up of the Great Frontier gave Europe an additional twenty million square miles, more than five times its existing area, and hence not just enormous additions to its wealth but also land and an immensely beneficial reduction in population density. The frontier thus represented a 'vast body of wealth without proprietors', an 'empty land', several times the size of Western Europe, a land 'whose resources had not yet been exploited'. Webb believed that this vast windfall acquisition

[16]   Ibid., p. 7.
[17]   Ibid., pp. 8–9.

generated a boom that lasted until the 1890s. It rescued Europe from its Malthusian trap and stimulated fresh thought and innovation.

This is not a view which many European historians have shared, preferring instead to stress continuity in Europe's own intellectual traditions and economic resources, questioning the financial benefits that New World bullion brought to the nations of Europe, and remaining sceptical about the value of the American contribution to European economy and culture. Indeed, part of the purpose behind Braudel's *The Mediterranean and the Mediterranean World in the Age of Philip II* (discussed in chapter 3) was precisely to question the excessive importance given to the emerging Atlantic economy and to re-establish the internal dynamics – environmental, economic and political – of the Mediterranean region. As Braudel pronounced in almost Gaullist tones in 1967, *'L'Amerique ne commande pas seule'*: America is not in sole command.

But, in the 1950s, this was far from being Webb's view. He found the manifestations of the frontier spirit 'present everywhere' in the societies created or affected by the Great Frontier – 'in democratic government, in boisterous politics, in exploitive agriculture, in mobility of population, in disregard for conventions, in rude manners, and in unbridled optimism'. Taking up a Turneresque theme, but giving it a new twist, Webb declared that while Europe could take the credit for mercantilism, capitalism and industrialism, 'the frontiers mothered democracy'. Its origins lay not in Europe, but 'in the forest glades of the New World', and especially along the eastern seaboard 'where Englishmen had come to make their homes'.[18]

Like Turner, Webb looked back on the closing of the frontier in the 1890s and lamented its passing. There were no new lands left: tropical Brazil and icy Alaska offered, in his view, poor prospects, nor did he place much confidence in the ability of science and technology to invent new frontiers through the more effective use of existing resources. He declared himself, like many

[18]   Ibid., pp. 5, 30.

other Americans, 'homesick' for the lost age of the Great Frontier and felt apprehensive about the future of Western civilization without it.

Even more emphatically than Turner, Webb saw the encounter with nature that the frontier represented as one of its most important characteristics. In Europe, a man's life was dominated by contact with his fellow men – in the form of laws, institutions and government. 'Everywhere he looked he saw that the great struggle was that of man against man. Nature, passive and recessive, had been pushed into the background by civilization.' By contrast, in the Americas, and on the other advanced lines of the Great Frontier, man confronted only nature and a vast, seemingly empty wilderness. 'What men had done to him all his life now fell away in a single instant: nowhere was there policeman, priest, or overlord to push him around.' There was little recognition here that these were already occupied territories with pre-existing cultures and land-use practices. They were 'virgin lands' once more. 'In Europe,' Webb concluded, 'the theme of life was man against man, man against civilization; but on the frontier the theme was man against nature.'[19] But such attempts to ignore the indigenes could not go unchallenged even in the history of the Americas.

'Indian Landscapes'

Webb, like his mentor, took little interest in the history of America before Columbus or even Jamestown. What mattered was the period of rapid change that followed thereafter. They were not alone in this. The common view in the nineteenth century and even after was that 'savage races have no history'. As one English ethnologist declared a quarter of a century before Turner unveiled his frontier thesis: 'each century sees them [the 'savage races'] in the same condition as the last, learning nothing, inventing nothing, improving nothing, living on in the same squalid misery

[19]   Ibid., pp. 31–2.

and brutal ignorance . . . They are without a past and without a future, doomed . . . to an inevitable extinction.'[20]

But one of the most substantial revisions – and refutations – of the Turner thesis has been to see the history of America from the other side of the ecological frontier, and to rescue the Amerindians from the state of nature and the lowly place of Darwinian savagery to which Turner, Webb and others assigned them. Both the narrow time-frame and the ethnocentrism of the Turner–Webb thesis have thus been radically challenged, and, with them, the ecological blindness. Michael Williams, for instance, has refuted the idea, so powerfully entrenched in Turner and many early writers on the American frontier, that America was an untouched land when Europeans first landed on the shores of North America. On the contrary,

> the Indians were a potent, if not crucial ecological factor in the distribution and composition of the forest. Their activities through millennia make the concept of 'natural vegetation' a difficult one to uphold. This does not mean that there was no untouched forest . . . but the idea of the forest as being in some pristine state of equilibrium with nature, awaiting the arrival of the transforming hand of the European, has been all too readily accepted as a comforting generalization and as a benchmark against which to measure all subsequent change. When the Europeans came to North America the forest had already been changed radically.[21]

Another writer makes a similar point. 'America did not begin,' he remarks, 'on the banks of the James River [in 1607], rather, when Asian peoples crossed the Bering Straits about 15,000 years ago.'[22] What Europeans took to be an empty land with a 'natural' landscape has now been shown to have been worked and modified

[20]   Frederic William Farrar, 'Aptitudes of races', *Transactions of the Ethnological Society*, 1867, in Michael D. Biddiss (ed.), *Images of Race*, Leicester, 1979, p. 148.
[21]   Michael Williams, *Americans and their Forests: a Historical Geography*, Cambridge, 1989, p. 49.
[22]   Karl W. Butzer, 'The Indian legacy in the American landscape', in Michael Conzen (ed.), *The Making of the American Landscape*, Boston, 1990, p. 27.

by Amerindians for thousands of years before the coming of the white man. There was a pre-European landscape that represented the errors and achievements of countless generations, and it is upon this cultural landscape that Euro-American patterns of land-use and settlement were later superimposed.[23]

This long-term, ecologically informed approach, which situates the frontier experience in the middle of American history rather than at its very beginning, is well represented in the work of two earlier writers, Carl O. Sauer and James C. Malin. In their different ways, these two scholars challenged many of the basic assumptions of the Turner thesis at a time when it still held a dominant position in American historiography, and, in so doing, helped to open up the wider pursuit of environmental history in the United States.

Working from a background in anthropology and geography (but with strong historical interests), Sauer sought from the 1920s onwards to reject the kind of environmental determinism Turner and his school represented, though he still continued to work with the concept of 'the frontier' that Turner had brought into prominence. Sauer rejected the idea that the first Europeans found a natural landscape in the Americas. 'Not even the more primitive non-agricultural folk,' he wrote in 1957, 'may be considered as merely passive occupants of particular niches in their forest environment.'[24] On the contrary, he saw landscapes as cultural products, fashioned to various degrees by human activity, and not least through the effects of fire and the domestication of crops and animals. Rather than a Darwinian sequence, advancing from primitive to civilized occupants and uses of the land, Sauer saw the landscape as a kind of palimpsest bearing the faded and half-erased signatures of several cultures in turn – Indian, Spanish, English and American.

He located the frontier not in some mystical interaction

---

[23] Ibid., p. 28.
[24] Carl O. Sauer, 'Man in the ecology of tropical America', in John Leighly (ed.), *Land and Life: a Selection from the Writings of Carl Ortwin Sauer*, Berkeley, 1963, p. 187.

between Western pioneers and pristine nature, but in the physical characteristics of the landscape, in vegetation and settlement patterns. The 'transforming hand of man' was first evident in 'Indian landscapes', even if the early Europeans were culturally blind to their existence or chose to ignore them in their desire to prove that this was 'free' land, unused, unoccupied, and so theirs for the taking. For Sauer, even more than in the passing references in Turner to 'Indian corn' and 'log cabins', the Europeans were the beneficiaries of their Amerindian predecessors, especially in the East where the American frontiersman 'took over the crops, methods, and fields of the Indian agriculturists'.

The perceived continuity between Amerindians and Europeans was all the greater for Sauer in that, unlike Turner, he paid particular attention to Mexico and Central America where Indian societies continued to exist in subordination to the Spanish rather than being destroyed or driven out by an advancing frontier as further north. For Sauer, the Spanish colonization of Mexico and the American south-west was a direct appropriation of indigenous peoples, and through them of the land, in such a way that the region retained aspects of its pre-conquest social and environmental characteristics.[25]

It was a theme of Carl Sauer's work – and that of many historians and geographers since – that what happened on the frontier was not the shedding of an obsolete European cultural skin and the reversion to, or rediscovery of, some kind of primitive mode of existence that lay closer to nature. Rather, imported values, which had been nurtured elsewhere, were mixed with, or laid on top of, those already in place and rooted in the American landscape. Even then there was no uniform response among the newcomers. When confronted with the American 'wilderness', different groups of Europeans responded in markedly different ways. Moreover, the frontier itself was not a single landscape, and thus unlikely of itself to produce a single immigrant response. The Spanish frontier missions in California were very different

[25]   Carl O. Sauer, 'Historical geography and the Western frontier' (1930), in ibid., pp. 47–8.

from the French fur-trappers and peasant settlers in Canada, or from the English, German and Scandinavian settlers moving out (each with their own chartacteristic style of farming, building, speech and dress) from their centres of diffusion (or 'cultural hearths') along the eastern American seaboard. The result was a complex mixture of nature and culture, as Sauer suggested in 1930: 'The nature of the cultural succession that was initiated in any frontier area was determined by the physical character of the country, by the civilization that was brought in, and by the moment of history that was involved.'[26]

So we are brought back to a rejection of a narrowly determinist approach to the environment and its effects on human history, and a re-statement of the importance of examining the long-term interdependence between culture and environment. At the same time, bringing the Amerindians (and other indigenous peoples elsewhere) back into history, and emphasizing their enduring cultural and ecological impact on the landscape, further diminishes the uniqueness Turner and others saw in the American experience and the expanding frontier.

James C. Malin, biologist and ecologist as well as trained historian, also refused to accept Turner's myopic time-scale and ethnocentric approach to the environment. He set the period of recorded history against the far longer backdrop of prehistoric and geological time. He also expanded the cultural parameters of the Turner thesis by seeing the landscape as offering various possible human uses. The physical environment of a region certainly set physical limits to what was possible, Malin believed, but within those limits people had considerable freedom of choice, and could employ their human ingenuity to overcome even some of its greatest apparent limitations.

Malin, like many other writers after Turner, stressed the great diversity of American ecologies: there was no single West, no uniform or typical frontier environment. For Turner, a native of Wisconsin, the frontier represented the forested mid-West and the Upper Mississippi Valley with which he was familiar; but for

[26]   Ibid., p. 49.

Malin the frontier was the very different environment of the semi-arid plains of Kansas. This was an area which northern Europeans, or those accustomed to the woodlands of the eastern seaboard, found hard to appreciate or know how to exploit. Malin quotes a telling passage from the account of Captain R. B. Marcy of an expedition in 1849 along what is today the Texas–New Mexico border. Marcy reported:

> Not a tree, shrub, or any other object, either animate or inanimate, relieved the dreary monotony of the prospect; it was a vast, illimitable expanse of desert prairie – the dreaded 'Llano Estacado' of New Mexico; or, in other words, the great Zahara of North America. It is a region almost as vast and trackless as the ocean – a land where no man, either savage or civilized, permanently abides . . . a treeless, desolate waste of uninhabited solitude.[27]

The region was not in any objective sense 'deficient' or 'inadequate', but it appeared so to those unfamiliar with such a seemingly empty and barren landscape. As a result, many early settlers avoided the plains entirely or congregated in the few places where trees and water were to be found. In time, however, Malin showed, the immigrant farmers found ways to make the land productive in their terms, principally through the use of artesian wells to bring water to the surface and the adoption of cereal crops able to thrive in semi-arid conditions. Malin concluded from this that when a group of people transferred their culture from one locality to another, they could not expect to be successful unless they utilized the plants and animals already present in the area or which adapted readily to local conditions. Nature did not predetermine a single cultural solution – white settlers might grow wheat where once Amerindians had hunted buffalo – but the environment did restrict the range of possible options. Civilization was thus a consequence of effective ecological adaptation and was 'successful' to the degree to which

[27] James C. Malin, 'Grassland, "treeless", and "subhumid": a discussion of some problems of the terminology of geography', in *History and Ecology: Studies of the Grassland*, Lincoln, Nebraska, 1984, p. 23.

immigrant farmers were able to harmonize their cultural inheritance with the local ecology.

From this 'possibilist' position, Malin contested another of Turner's environmentalist assumptions: the pessimistic idea of 'closed space' created by the ending of the frontier and loss of 'free' land in the 1890s. In Malin's view, an area like the Kansas plains was never entirely 'closed'. It was always open to new cultural possibilities, for the land to be used in a number of different ways. What was useless in the eyes of one culture might be the valued resource of another. Thus did Malin shun the Malthusian scenario.

7
The Environmental Revolution

A 'Pivotal Event'

But, one might ask, was not Turner at least partly right in seeing the frontier as a uniquely transformative process? In environmental, as much as in cultural terms, was not America a changed place once the frontier of expansion had rolled westward across the land? Three American historians have remarked that whatever the 'contradictions and errors of his scholarship', Turner was 'surely right' to see the long European (and, they quickly add, African and Asian) invasion of North America, and the resistance to it by the continent's existing inhabitants, as 'the pivotal event in American history'.[1]

The research of Sauer, Malin and others has helped to establish the long ancestry, the cultured forms and the ecological continuities of the American landscape, and in so strongly aligning history with anthropology, archaeology and ecology, such scholars have helped to break through the cultural barrier of writers like Turner who saw North America as empty of people and empty of history before the arrival of the westward-moving frontier. Although apparently unaware of the parallel, Sauer and Malin brought

---

[1] William Cronon, George Miles and Jay Gatlin, 'Becoming West: towards a new meaning for Western history', in W. Cronon, G. Miles and J. Gatlin (eds), *Under an Open Sky: Rethinking America's Western Past*, New York, 1992, p. 6.

American historiography more closely into line with the Annalistes in France, with their emphasis on the creative partnership of history and geography, the physical structure beneath human history, and the underlying continuities of a much-extended North American *longue durée*.

At the same time, the very process of recovering the pre-Columbian past, and retrieving the 'Indian landscape' from neglect and obscurity, has in some respects made the sense of discontinuity and disjuncture even more profound and distinctive from the less-interrupted flow of Europe's own environmental history. Sauer was not blind to this paradox of continuity and disjuncture. He recognized, for instance, that the 'personality' of Mexico, like that of much of Hispanic America, had been deeply affected by two factors. One was the 'deep, rich', pre-Columbian past and the living 'continuity with ages long gone'; the other was the abrupt but enduring impact of the radical changes introduced by Mexico's Spanish invaders in the sixteenth century.[2]

Historians of North America remain impressed by the profundity of the ecological *and* cultural transformation that accompanied the advent of the Europeans. In a study of the trans-Appalachian frontier in the late eighteenth and early nineteenth centuries, Malcolm J. Rohrbough commented (still in part under Turner's influence) on the great changes affecting the Ohio Valley following the arrival of the first European settlers:

> Indians had lived on these lands for a thousand years and changed them little. Now, within one generation, the American pioneer would transform them. The Great Migration had many of the qualities of a revolution: for the Native Americans thrust aside; for the pioneers who took the land and exploited it; for the land itself.[3]

[2]   Carl O. Sauer, 'The personality of Mexico' (1941), in John Leighly (ed.), *Land and Life: a Selection from the Writings of Carl Ortwin Sauer*, Berkeley, 1963, p. 105.
[3]   Malcolm J. Rohrbough, *The Trans-Appalachian Frontier: Peoples, Societies, and Institutions, 1775–1850*, New York, 1978, p. 158.

The character and consequences of that 'revolution' – involving indigenous peoples, European colonists and ecology – are something recent historiography has sought to capture (and in the history of Africa and Australasia as well as the Americas). If Turner's preoccupation was with what nature had done for the frontiersman – giving him democracy, individualism and a distinctively American identity – today's emphasis is upon what the frontiersman and the colonist did to the Indians and the land or how whites and native Americans represented radically different approaches to and uses of the same contested landscape.

In a sensitive study of colonial New England which epitomizes the general trend of much recent scholarship in environmental history outside Europe, William Cronon observes that, while it might be tempting to believe that when the Europeans first arrived in the Americas they confronted a 'virgin land', in reality: 'Nothing could be further from the truth. In Francis Jennings's telling phrase, the land was less virgin than it was widowed. Indians had lived on the continent for thousands of years, and had to a significant extent modified its environment to their purposes.'[4]

Cronon demonstrates Indian interaction and modification of the environment in a number of ways. For instance, the Indians used fire to clear old undergrowth and encourage the growth of new grass and green shoots, which in turn would attract game animals (a practice also followed by the Aborigines in Australia, another supposedly empty land). He shows how in New England the Indians followed a way of life that required an intimate acquaintance with the local ecology, knowing how to survive the different seasons by knowing when to fish, when to hunt game, or sow and harvest field crops. Their way of life was thus geared to the diversity and seasonal rhythms of the ecology, while at the same time imposing certain modifications upon it – through the use of fire, by making clearings in the forest, and by selecting certain plants for cultivation. When the first colonists arrived in

---

[4]   William Cronon, *Changes in the Land: Indians, Colonists, and the Ecology of New England*, New York, 1983, p. 12.

the seventeenth century they had little sense of the Indians' skill in exploiting the environment to meet their subsistence needs, and wondered at their apparent poverty in a land which, after England, appeared filled with all the abundance of a provident nature. However, the colonists themselves proved initially inept at trying to draw a living from the New England landscape, especially in winter. Only gradually were they able to establish themselves, adapt to the new conditions, and then impose their own system of land-use, based upon English farming methods and notions of property, and so to oust the Indians and replace their use of the land. Cronon comments: 'The destruction of Indian communities . . . brought some of the most important ecological changes which followed the Europeans' arrival in America. The choice is not between two landscapes, one with and one without human influence; it is between two human ways of living, two ways of belonging to an ecosystem.'[5]

While historians remain critical of Turner's undifferentiated and Darwinian understanding of the frontier and the changes it wrought, they continue to see the advance of the ecological frontier as a series of stages, not all present everywhere in the same order or importance, but building cumulatively into a profound and relatively rapid process of environmental and cultural transformation. And while the sequence has perhaps been most extensively analysed in the context of the North American frontier, the model is one – as Crosby, among others, has shown – which appears to fit other parts of the Americas (including Canada, the West Indies and Latin America) and, in certain respects, South Africa, Australia and New Zealand as well.

Furs and Forests

One of the most immediate and momentous of these ecological changes – disease – has already been discussed in chapter 5, but it is important to remember the continuing role of disease in

[5]   Ibid.

physically and culturally undermining indigenous societies and their ability to resist or adjust to European invasion. A second stage in the advance of the ecological frontier, especially in Canada and the American North-west, but with close parallels elsewhere by sea and land, was the fur trade. Europe's demand for furs seemed insatiable, for warmth, for fashion and for status. By the seventeenth century, Europe had largely exhausted its own supply of animal furs, driving a number of fur-bearing animals close to extinction. Only in Siberia was it still possible to trap such animals in significant numbers (one reason why trade with Muscovy was much sought after by English merchants in Tudor times). The opening up of Canada by the French and English via Hudson Bay, the St Lawrence River and the Great Lakes held out the prospect of a lucrative and seemingly inexhaustible new source of furs, especially much prized beaver skins. The resulting trade was massive – and horrific. In just one year, 1742, Fort York handled 130,000 beaver skins and 9,000 marten pelts; in the 1760s, one of the posts of the Hudson's Bay Company supplied nearly 100,000 beaver skins. The size of the slaughter can be judged, too, from the volume of furs reaching Europe. In 1743 the single French port of La Rochelle imported 127,000 beaver skins, 30,000 martens, 12,000 otters, 110,000 racoons, and 16,000 bears.[6]

In the face of such wholesale slaughter, many fur-bearing animals quickly became extinct in the more easterly parts of North America and fur traders and trappers had to push further and further west in pursuit of their prey. By 1831 the beaver was extinct in the northern Great Plains, and by the 1830s only 2,000 skins a year could be collected from the Rockies. By the 1820s and 1830s, after barely two hundred years of hunting and trapping, North America's fur trade was effectively over. There were few animals left to trap, and a whole rich spectrum of American wildlife had disappeared.

The fur trade has been much discussed by historians for it intersects with several seemingly different narratives. In political

[6] Clive Ponting, *A Green History of the World*, Harmondsworth, 1992, p. 181.

terms, competition for furs helped spark several rounds of Anglo-French conflict in North America. The commercial lure of the trade was also a stimulus to geographical exploration and close contact by trappers, traders and Jesuit missionaries with the Indians of the Great Lakes and beyond. But, from our perspective, it is also a prime example of the way in which the ecology of the Americas was transformed not by the invasion of Old World plants and pathogens, but by the deliberate human plundering of a once rich environment. It is an illustration, as vivid as that of sugar plantations in the tropics, of how Europe's remorseless quest for profit could transform, even destroy, distant environments, which the great majority of consumers in Europe would never see and perhaps not even pause to consider. This was the imperialism of greed and opportunity, not of nature.

A further stage in the expansion of the ecological frontier, especially in North America, and even more massive in its environmental impact, was the destruction of forests. This, too, was an act of human appropriation and displacement, resulting in the expulsion of the Indians who had hitherto managed the forest eco-system according to their own needs. With the destruction of the trees went the material means that had sustained their livelihood as well as the animals for whom the forests had provided sustenance and shelter. Like the fur trade, deforestation turned nature into a commodity, timber for ships' masts and spars, hardwoods like mahogany for furniture-making. Europe, profligate with its own woodlands, found in America a seemingly ceaseless sawmill able to keep its ships afloat and its drawing rooms supplied with elegant tables, chairs and writing desks. In America, timber was so cheap and plentiful it was used for fuel, for construction purposes and almost every conceivable item of everyday use. In many parts of the eastern seaboard and the mid-West trees were seen as an obstacle to be cleared as quickly as possible in order to create farming land. The visual as well as ecological transformation could not have been more striking, nor more in contrast with the slow and fitful progress of deforestation in Europe in the centuries before the Black Death.

'Other than the creation of cities,' Michael Williams observes, 'possibly the greatest single factor in the evolution of the American landscape has been the clearing of the forests that covered nearly half of the country.'[7] By his reckoning, 45 per cent of the area now occupied by the United States was wooded in the seventeenth century. By 1920 that figure had been halved, with much of the initial destruction – some 300 million acres of forest – completed by the early twentieth century. It was as a witness to this rapid and destructive process of deforestation that George Perkins Marsh (as noted in chapter 3) was prompted to make his warning comments in *Man and Nature* in the 1860s. For Marsh, as for many other Americans since, deforestation has been the single most momentous and significant aspect of environmental change.

The West Indies Transformed

The importance of deforestation and its accompanying ecological and social changes was by no means confined to the United States. In some ways it was even more striking in the more fragile island eco-systems of the West Indies, as described by Carl Sauer and, more recently, by David Watts. Before the arrival of the Spanish in 1492, the indigenous population of the islands, the Arawaks and Caribs, lived mainly by clearing patches of forest to grow root crops such as cassava and sweet potatoes on raised beds, known as *conucos*. Their diet was supplemented by seasonal fishing and fruit-gathering. The island Indians thus made full use of what Watts calls a 'beneficent but sometimes difficult environment'. Not only was their diet relatively nutritious, but (despite the improbably large numbers attributed to them by Cook and Borah) their impact on the ecology of the islands was likely to have been relatively modest, certainly when compared with the depredation that was to follow, though it is important to avoid the temptation to understate the degree of change taking place in the ecologies of the pre-Columbian Caribbean in order to make

---

[7] Michael Williams, *Americans and their Forests: a Historical Geography*, Cambridge, 1989, p. xvii.

the subsequent process of destruction seem all the more momentous. Watts is surely tempted towards an over-idealized vision of this sort when he remarks of the Arawaks: 'It appears that they lived largely in harmony with themselves and with their environment.'[8]

Quite apart from the continuing processes of forest destruction and regeneration caused by hurricanes and tropical storms, the Indians themselves relied upon localized destruction in order to grow their crops. Burning vegetation to clear sites for cultivation was important as a way of releasing nutrients which were locked up in the vegetation into the soil. Like the Indians of the North American mainland, the Arawaks often grew several crops together on their *conucos*, thereby maximizing the use of nutrients and minimizing the dangers of soil erosion from heavy rain. As their fertility declined after a few seasons, the cultivated areas were abandoned, and left fallow for thirty years or so, allowing secondary forest to grow and saving the soil from leaching and erosion.

The arrival of the Spanish, followed by other Europeans, brought fundamental changes to this Indian way of life. The actual duration of the conquest of the Caribbean was brief, occupying the years between 1492 and 1519, but, like a hurricane, it was devastating. The Spanish pillaged the islands in their quest for gold and slaves; imported diseases were rampant; and cruelty and forced labour further reduced the population. Then, having stripped the islands of their most accessible forms of wealth, the Spanish settlers and adventurers turned their attention to the American mainland, with Cortes's expedition to Mexico in 1519. But the effects of the Spanish invasion were long lasting and profound and allowed no opportunity for the old Amerindian culture and island eco-systems to re-establish themselves.

The first inroads of diseases were quickly followed by 'species shifting' with the importation of new plants and animals. Pigs

[8] David Watts, *The West Indies: Patterns of Development, Culture and Environmental Change since 1492*, Cambridge, 1987, p. 53.

brought a new threat to Indians' survival, as they devoured crops, multiplied rapidly and ran wild in the woods. Bartolomé de las Casas, who first arrived in Hispaniola in 1502 and was distressed at the wholesale destruction of the pre-conquest way of life, wrote that 'there were great numbers of pigs feeding on sweet roots and delicate fruit . . . the hills were full of them.'[9] Eventually, in the seventeenth century, long after the Indians themselves had been driven to extinction, the feral pigs were brought under control: Europeans shot them for food or for sport. Although the Spanish were unable to establish in the Caribbean some of the most prized items of their Mediterranean biota, such as wheat and sheep (though the latter flourished elsewhere, in Mexico and Chile), bananas were successfully introduced from the Canary Islands and cattle, first introduced by Columbus on his second voyage of 1493, spread across the islands, adding to the destructive impact of the pigs, their hooves compacting the exposed ground, causing rapid run-off and severe soil erosion.

Initially, Europeans had much to learn from the island Indians, as they did from those of New England and Virginia. Their initial survival depended on food produced by the *conuco* method, just as they learned how to kill hardwood trees by ringbarking and burning them rather than by the laborious process of trying to cut them down with quickly blunted axes. They learned how to grow cassava, how to sleep in hammocks, and make liquor from fermented sweet potatoes (until, with rising sugar production, they could get drunk on cheap rum). But, increasingly, the Indians were pushed aside, their system of land-use discarded and obliterated from the landscape. By the early seventeenth century, few visible signs of the *conucos* remained. The colonists turned to tobacco and then sugar, which did more than anything else to transform the ecology of the West Indian islands.

The development of sugar plantations based upon slave labour was clearly motivated by a quest for profit. Encouraged by Dutch capital and the introduction of new production methods from Brazil, sugar spread rapidly through the islands of the West Indies

[9]   Quoted in ibid., p. 117.

from the 1640s onwards. By the late seventeenth century, sugar imports exceeded the combined value of all England's other colonial produce, and such was the demand for sugar in Europe that for much of the eighteenth century West Indian plantations produced a quarter of all British imports. The rise of sugar as a plantation crop had three closely interrelated environmental effects: it hastened the destruction of what remained of the pre-Columbian island eco-systems; it was accompanied by the mass importation of African slaves and the elimination or marginalization of Amerindian and poor white populations; and it heralded the introduction of new plants (above all sugar itself) and new diseases (especially, as we saw in chapter 5, yellow fever). On each of these fronts, sugar represented a significant advance of the ecological frontier.

The environmental effects of the sugar revolution were nowhere more evident than on Barbados, where by the mid-seventeenth century the last remaining tracts of pre-Columbian forest were cleared to make room for sugar plantations. An island once thick with trees became an open landscape given over almost entirely to sugar-cane. So little of the original tree cover remained, even in relatively inaccessible areas, that the chance of regeneration by native flora and fauna was virtually eliminated. Indigenous plants, birds and animals died out, and even those few that somehow survived were extirpated as 'weeds' and 'pests'. Stripped of the tree cover and the nutrients it provided and subject to the relentless demands of the sugar crop itself, the soil became impoverished and was washed away through erosion and gullying. Although the speed and extent of environmental change on Barbados represents an extreme case, even in the West Indies, its history of species loss and replacement, of the total transformation of the landscape and the elimination of previous cultural traces, provides a clear and well-documented example of how comprehensive the process of environmental change could be. In the West Indies, as elsewhere in the tropics, plantations were one of the most powerful of all agents of environmental change. The case of Barbados also highlights the importance of slavery, of control over people as much as places, to the history of environmental change in the Americas, Africa, Asia and the Pacific since the early sixteenth century.

'A Howling Wilderness'

What was the driving force behind the expansion of the ecological frontier? Three broad, but not necessarily mutually exclusive, explanations have been offered. One, as Crosby and McNeill have argued, is that it was essentially biologically driven. Rather than a conflict of cultures, there was a clash of ecologies, as Old World diseases, plants, animals, and with them people, combined to exploit biologically favourable conditions and establish themselves at the expense of indigenous rivals. But, as has already been suggested, this seems an incomplete answer. In many instances, the success of the biological depended on the cultural, on the ability or determination of the Europeans – as in the case of the plantations of the West Indies – to reshape economic, social and political conditions to suit their commercial interests. Amerindian societies were not destroyed by smallpox alone, however virulent it may have been, nor by pigs and cattle, destructive though they could undoubtedly be to pre-existing systems of land-use, but by the imposition of a completely new way of life and a new way of exploiting and refashioning the environment.

The second possibility, which has already been touched upon in passing but which is evident enough in much of the literature of European expansionism and the expanding frontier, was the driving force of capitalism supported by its myriad scientific and technical agencies. In this sense, the environmental history of the Americas was but an extension of the economic history of Europe, where capitalism had long since taken over from feudal lords and monastic orders the responsibility for transforming 'waste' land into productive agriculture and turning idle nature into commercial profit. The spirit of capitalism revealed itself in the Americas in as many diverse human forms as it did in Europe, but the perception of the New World as a land of opportunity and wealth, property and profit, was a strand common to them all and distinguished the invaders and immigrants from the native Americans they subordinated, expelled or exterminated. Ultimately, then, it can be argued, economics rather than biology determined the post-conquest environmental history of the Americas.

There is, however, a third possibility that merits consideration. This is the idea, rather at variance with both Turner and Crosby, that the expansion of the ecological frontier was a product of cultural determinism, of values derived directly from Europe which, through their transference to the New World, had a transforming effect on the American frontier. This explanation, to be sure, need not necessarily be divorced from the economic one, though in giving priority to culture over commerce it is often, in effect, given an autonomy of its own.

In one of its more concise forms, this interpretation has been used to explain the specific character of the Spanish conquest of Mexico and Peru. The year of Columbus's landfall in the Americas – 1492 – was also, with the fall of Granada, the year of the final stage of the reconquest of the Iberian peninsula from the Moors, and the two have logically been linked. The Spanish carried with them to the Americas a crusading spirit, the confidence born of believing that they were the chosen people of God fighting against unbelievers and sustained by the myths and legends of medieval Christianity. To many *conquistadores*, 'the conquest of America was simply an extension of the Reconquest in the peninsula.'[10]

Accustomed in Spain to a military frontier, bristling with armed skirmishes, the conversion of the 'heathen', and colonization through the creation of towns, Cortes and his successors followed this familiar pattern in the New World as well. More than that, Spain was exceptional in Europe at the time in being so attached to a pastoral economy. Large areas of the provinces of Estramadura and Andalusia from which so many of the *conquistadores* and their followers came were dominated by huge flocks of sheep and

[10]   Alistair Hennessy, *The Frontier in Latin American History*, London, 1978, p. 28.

vast herds of cattle. It was this pastoral economy that the Spanish transferred to the Americas in 1493, with the devastating social and environmental effects already touched upon. Moreover, the culture of pastoralism, restless and rootless, was so intrinsically alien to the Azetcs and Incas, with their unfenced fields and vulnerable crops, that the 'earth-bound' sedentary societies of the Amerindians were rapidly pounded into the dust. It was not the cattle, pigs and sheep themselves, as Crosby would lead us to imagine, but the very different cultural system they represented that may be said to account for the rapid collapse of the Amerindian societies of Central and South America.

This is a specific case, but the cultural dynamic behind the advance of the ecological frontier has been identified, much more generally, with European civilization and the Christian tradition and not just with the fervent, restless Catholicism of the Spanish *conquistadores*. In 1967, at a time when the ecology movement in the United States was beginning to gain momentum following the publication of Rachel Carson's *Silent Spring*, Lynn White wrote a provocative and influential article on 'the historical roots' of the current environmental crisis. He argued that Western views of the natural world had been profoundly influenced by a Judaeo-Christian tradition, which sees God as entrusting dominion over nature to humankind. White (and others following a similar line of argument) find the origins of this in Genesis: 'Be fruitful and multiply, and fill the earth and subdue it; and have dominion over the fish of the sea and over the birds of the air and over every living thing that moves upon the earth . . . Be fruitful and multiply and replenish the earth and subdue it.'

On the basis of such sentiments, White believed that the figure of St Francis of Assisi was quite exceptional in the Christian tradition. Far from a love of birds, plants and animals, Christianity appeared to him to be 'the most anthropocentric religion the world has seen', a religion that showed unparalleled 'arrogance towards nature'. It had deliberately set out to destroy pagan religious beliefs which had served to protect trees, animals and the land itself from ruthless human exploitation and destruction. 'By destroying pagan animism,' White averred, 'Christianity made it possible to exploit nature in a mood of indifference to the feelings of natural subjects.' He concluded that, 'despite Darwin, we are *not*, in our hearts, part of the natural process. We are superior to nature, contemptuous of it, willing to use it for our slightest whim.'[11]

White's claims have drawn a powerful and often emotional response from many environmentalists and have encouraged a critical reappraisal of attitudes to nature in the Jewish and Christian traditions and a willingness to look more favourably upon the green credentials of other faiths. But not everyone has found it possible to agree with White's sweeping claims or found that his ideas stand up to closer historical scrutiny. It has reasonably been doubted whether Christianity is in any sense uniformly hostile to nature: it has been pointed out, for instance, that there was a strand of medieval Christianity that glorified God in nature and saw in it evidence of His divine work, or that stressed human stewardship rather than mastery over the natural world. It was perhaps only in the seventeenth and eighteenth centuries that these ideas became lost, overwhelmed by mounting economic, scientific and technological change and by the rapacious spirit of profit-seeking capitalism. Equally, although the idea that Buddhism, Hinduism and other non-Western religions are far more environmentally friendly than Christianity has gained ground and become part of an environmentalist critique of Western civilization in many areas of the world, the reality behind the rhetoric is often unconvincing. There is a massive discrepancy between what religious texts enjoin and what hard-pressed peasant societies actually practise, and the 'nature' that is extolled and venerated in Asian cultures is not 'nature in the raw' but of the highly selective and anthropocentric kind favoured by the poet, the painter, the physician and the maker of gardens and palaces. None the less, White's essay struck a note which has echoed through much subsequent environmentalist writing, and helped to establish the idea, *contra* Turner, that Europeans carried with

[11]   Lynn White, 'The historical roots of our ecologic crisis', in *Machina Ex Deo: Essays in the Dynamism of Western Culture*, Cambridge, Mass., 1968, pp. 75–94.

them well-established environmental ideals and prejudices that helped transform an alien landscape.

In a far-reaching, if excessively one-dimensional, study along these lines, Frederick Turner (not to be confused with the Frederick Jackson Turner of the 1890s) argued that the cultural, or, as he would prefer it, the spiritual, dynamic behind Western attitudes to nature can be traced back to the early history of the Middle East and the antipathy that there developed between civilization and cities, on the one hand and, on the other, the desert, the wilderness and the hostile wildness of nature. It was there, Turner argues, that 'human beings began to enact the dream of mastering the natural world', a dream that later became incorporated into the Old Testament and hence into Christian tradition. The desert wilderness was, as Moses perceived it, a 'great and terrible' place, and it was the task of man to subdue it. Such ideas, 'codified and embedded in Scripture', travelled to the New World with the Pilgrim Fathers, and helped shape the destructive impulses of the early colonists and their deep antipathy to the forest 'wilderness' they encountered.[12]

The Judaeo-Christian tradition and the ancient Middle East is not the only place where American authors have imaginatively sought to locate the origins of New World attitudes to nature and the wilderness. Roderick Nash, in a highly regarded work first published in 1967, argued that such negative environmental attitudes were deeply rooted in the mind of Western man, mainly through the experience of living for centuries in the dark and almost endless forests of northern Europe. The fear and repulsion generated by these forests, with their associations with evil, lawlessness, strange events and menacing inhabitants (it is ironic that we are led back to precisely the Germanic folk and forests Frederick Jackson Turner was so eager to leave behind!), created an enduring reaction against the wilderness. The very term 'wilderness', Nash points out, was an invocation of wildness and bewilderment: it represented 'the unknown, the disordered, the

---

[12]   Frederick Turner, *Beyond Geography: the Western Spirit against the Wilderness*, New Brunswick, 1983, pp. 21–2, 41.

uncontrolled' in humans as in nature, and a 'large portion of the energies of early civilizations was directed at defeating the wilderness in nature and controlling it in human nature.'[13] By contrast with the dark and menacing forests of the north, Europe, from the time of the Greeks, favoured a pastoral idyll, a cultivated, sun-lit and orderly landscape of the kind celebrated by Lucretius, Horace and Virgil. The Bible reinforced this ideal for the Christian West, for 'anyone with a Bible had available an extended lesson in the meaning of wild land.'[14] In medieval and early modern times, the wilderness of the Bible, sun-scorched, stoney and arid, became confused with the 'fearsome' and 'doubtful' land the *Beowulf* saga depicts, a dark and 'howling' wilderness of trees, bogs and supernatural beings, just as the Biblical Eden had become mixed up with the Persian and Islamic idea of a paradise garden.

This perceived dichotomy between nature as either wilderness or paradise, Nash argued, crossed the Atlantic with the first European colonists. William Bradford, arriving on the *Mayflower* in 1620 was but one of many who found in the New World not the Eden of his expectations but a 'hideous and desolate wilderness'. 'There was,' Nash observes dryly, 'too much wilderness for appreciation.'[15] Old fears and ancient insecurities were revived in face of such a vast, forbidding wilderness: there was a fear of losing hold of civilization entirely, of becoming wild, unruly, and ungodly. Michael Wigglesworth stated in 1662 that, outside the few settlements that then existed in New England, there was nothing but

> A waste and howling wilderness.
> Where none inhabited
> But hellish fiends, and brutish men
> That Devils worshipped.[16]

[13] Roderick Nash, *Wilderness and the American Mind*, 3rd edn, New Haven, 1982, p. xi.
[14] Ibid., p. 8.
[15] Ibid., pp. xii, 23–4.
[16] Quoted in Hans Huth, *Nature and the American: Three Centuries of Changing Attitudes*, 2nd edn, Lincoln, Nebraska, 1990, p. 6.

The first Europeans regarded the American wilderness as much as a moral as a physical wasteland, crying out for Christianity and conquest. In the minds of the Puritans, this Biblical wilderness was a test of their faith and endurance; theirs was the divinely ordained task of transforming the 'black' and 'howling' wilderness into an orderly, earthly Eden. Such attitudes continued, Nash argued, to dominate American responses to the wilderness until the latter half of the nineteenth century. Only then, with the closing of the frontier in the 1890s, did American attitudes change and a new concern for conserving nature begin to emerge through the national parks movement. This eventual transition, from destruction to conservation, was 'nothing less than revolutionary', for, according to Nash, it went against the entire history of Western man's 'ancient biases against the wild'. By contrast with that long tradition, 'today's appreciation of wilderness represents one of the most remarkable intellectual revolutions in the history of human thought about land.'[17]

Whether or not we accept the arguments offered by post-Turnerian writers about the ancient ancestry of American ideas about nature and the wilderness (and many of them seem unduly simplistic), it is surely necessary to recognize the importance of cultural rather than purely economic or ecological arguments about Western environmental attitudes. One of the ways in which we can do this is by returning to the idea of the landscape as a cultural rather than purely physical construction. The ecological frontier was also, it can be argued, a frontier of the mind, a cultural landscape in the making.

'Nature's Continent'

It is possible to use the term 'cultural landscape' in two rather different ways. When Sauer used the expression in the 1930s in connection with the American frontier, he meant a physical landscape which had been subjected over time to a succession of

[17] Nash, *Wilderness*, p. xii.

different cultural presences, each of which – Indian, Hispanic, Anglo-American – left its imprint upon the land, perhaps in the form of burial mounds, or the ancient trackways followed by modern roads and railways, or the kinds of plants that grew or animals that grazed. The landscape thus represented the accumulation of several overlaid cultural traces which together made up the visible landscape. W. G. Hoskins approached the 'making of the English landscape' in the 1950s in a not dissimilar fashion: his was a study of the 'historical evolution' of the landscape as a 'man-made' rather than purely natural 'creation'. But historians, geographers and others are at least as likely these days to think of a landscape as something which is 'invented' or 'imagined' rather than physically constructed, a cultural artefact produced by the artistic, scientific or political imagination rather than an actual physical entity. The landscape exists not so much 'out there', to be toiled and trampled over, as in the minds of those who see it and interpret it, perhaps merely write about it from a convenient distance, and endow it with associations, symbolic connections and metaphorical meanings that are of their own invention.

It is sometimes argued, in this way, that America was not 'discovered', whether by Columbus or anybody else, but had to be 'invented'. Columbus had been in search of the spice islands and China when he stumbled upon America's shores. He had little sense, even after several voyages of exploration, that he had arrived at a continent hitherto unknown to Europe. The geographical outlines of America were only gradually pieced together through exploration; and islands, lakes and mountains had to be brought into European consciousness by being named and located on maps. Only thus was the identity of America as a continent to rank alongside Europe, Asia and Africa fully determined. But beyond its geographical form, the cultural identity of this New World had yet to be established. Over the centuries, the philosophers and historians of Europe, from Hippocrates and Herodotus to Montesquieu and Hegel, had given Europe, Asia and Africa their respective meanings and identities. But what did this New World of America signify?

As we will see more clearly when we come to consider Alexander von Humboldt in chapter 8, the environment came to be seen as an essential part of what made America distinctive from its continental cousins. The vastness of the landscape and the grandeur of its natural features, the seemingly limitless abundance of its wildlife, but also the apparently primitive character of its indigenous human inhabitants – these seemed to be America's principal features. This was 'nature's continent', and Frederick Jackson Turner in the 1890s was, unconsciously perhaps, following in the tradition of Montaigne in the seventeenth century and Rousseau in the eighteenth in believing that there was a singular correspondence in the Americas between people and place. For Montaigne and Rousseau the way of life of the Indian, at least in the abstract and from a distance, was to be admired and contrasted with the constraints and artificiality of European civilization. For Turner it was a primitive life, closer to raw nature than civilized society, but it needed to be learned from by the pioneer and frontiersman. But that nature was a vital, even dominant, presence in the Americas, scarce anyone denied.

Landscape and National Identity

The importance of an appreciation of one's own country through its natural sights and wonders as well as its representative plants, birds and animals (rather than the mythical beasts of the medieval world) began to form a significant basis for emerging national identities in Europe as early as the sixteenth century. By the late eighteenth century in England there was a growing sense of pride in this 'green and pleasant land'. A similar sense of emotional and ideological identification was evident in the landscape painting of Constable and Turner and, more intimately, in Samuel Palmer and the work of countless watercolour artists of the period. At a time when Wordsworth and the Romantic poets were nurturing a more passionate involvement with nature and the English countryside, the French Revolution and the Napoleonic wars restricted artists' access to the landscapes of Italy, rife with classical literary and historical allusions, and encouraged them to reflect upon the countryside nearer to home. A taste for 'sublime' or 'picturesque' views of nature could be met by scenes of Lakeland beauty or the celebration of an English rural landscape as yet largely unsullied by urban sprawl and the industrial revolution.

Similar invocations of landscape as identity found expression across the Atlantic, where the celebration of nature became an important mark of American self-esteem and emerging national pride. This is a history that has often been traced, and credit has rightly been given to the aesthetic and scientific contribution made by Thomas Jefferson, firstly through his *Notes on the State of Virginia*, originally published in 1781, and, secondly, through his appointment while President in 1804 of the Lewis and Clark expedition to the American North-West. The ideal of the 'Jeffersonian landscape' placed land at the centre of American life and ideology, fostering the view of the United States as a nation of yeoman farmers and the country as 'a well-ordered green garden magnified to continental size'.[18] Jefferson also invoked the natural landscape of the New World as an integral part of American spirit and pride, doubting (in the face of European sceptics like Buffon) that it lost anything by comparison with the famed and more familiar landscapes of Europe. There was 'sublime' delight and 'rapture', he claimed, in viewing Virginia's Natural Bridge, 'so beautiful an arch, so elevated, so light and springing as it were up to heaven'. Or in the spot where the Potomac River broke through the Blue Ridge near Harper's Ferry to join the Shenandoah, and presented 'perhaps one of the most stupendous scenes in nature'. It was a scene, Jefferson could not resist adding, that was 'worth a voyage across the Atlantic'.[19]

Considerable though his personal and political influence undoubtedly was, Jefferson was not alone in expressing enthusiasm for America's landscape and natural history. The landscape painters of the Hudson River School in the 1820s expressed a

[18]   Leo Marx, *The Machine in the Garden: Technology and the Pastoral Ideal in America*, New York, 1964, p. 141.
[19]   Quoted in Huth, *Nature and the American*, pp. 19–20.

new Romantic sensitivity to American scenery, finding in it suitable artistic and patriotic subject matter despite the lack of historical and mythological associations. The 'wilderness' was beginning to emerge in the minds of many Americans as something unique, with no counterpart in the Old World, and a source of manliness and pride. William Cullen Bryant, a most ardent believer in the distinctive appeal of the American wilderness, expressed his disappointment on visiting Italy in 1834, commenting that 'if the hand of man had done something to embellish the scenery, it has done more to deform it . . . the simplicity of natural scenery . . . is destroyed.'[20] A year later, in an 'Essay on American Scenery', Thomas Cole, one of the leading painters of the Hudson River School, contrasted the scenery of Rome with that of the American West. There, too, he argued it was possible to feel 'the emotion of the sublime', but, unlike Italy, its landscape rich in classical allusions and littered with the ruins of an ancient civilization, in the West it was 'the sublimity of a shoreless ocean unislanded by the records of man'.[21]

Identification with the grandeur and untouched beauty of a landscape that was still close to raw nature was also evident in the novels of James Fenimore Cooper and the verse of Walt Whitman. Cooper's novel, *The Pioneers*, published in 1822, provided a virtually Jeffersonian account of a society of yeoman farmers 'who respected nature even while they transformed it'. For Cooper, the American landscape was not the 'howling wilderness' of the Pilgrim Fathers, threatening, loathsome and ugly, but a place of wonder, beauty and adventure. Anticipating Henry David Thoreau and *Walden* by thirty years, Cooper found a moral influence in 'the honesty of the woods' and a rich source of self-discovery.[22] This visual and moral delight in a distinctly American landscape extended steadily westward, an advancing cultural frontier of aesthetic appreciation and national pride, to embrace Niagara,

[20]   Quoted in ibid., p. 33.
[21]   Quoted in Katherine Emma Manthorne, *Tropical Renaissance: North American Artists Exploring Latin America, 1839–1879*, Washington, 1989, p. 93.
[22]   David C. Miller, *Dark Eden: the Swamp in Nineteenth-century American Culture*, Cambridge, 1989, p. 136; Nash, *Wilderness*, p. 76.

8
Inventing Tropicality

'Otherness'

Recent scholarship has often attempted to conceptualize the relationship between Europe (or the West) and other parts of the world in terms of a principle of 'otherness'. In his seminal study of *Orientalism*, Edward W. Said argued that the Orient was not a geo-political 'fact', but a political and cultural creation of the West. Orientalism was 'a system of representations framed by a whole set of forces that brought the Orient into Western learning, Western consciousness, and later, Western empire'. It was (and, for Said, has continued to be) a way of seeing, thinking about and representing the Orient, characterizing it in terms of certain stereotypical features, defining it in ways which simultaneously express both difference from the West and inferiority to it.[1]

Interpretations of otherness have tended to focus upon representations of non-Western people and their cultures rather than the otherness of non-European environments. It can be argued, however, that alien landscapes were often imbued with as much importance as peoples or cultures themselves. Landscapes, and other aspects of the physical environment, such as climate and disease, were endowed with great moral significance. Environments, as we have seen, were widely believed to have a determin-

[1] Edward W. Said, *Orientalism*, London, 1978, pp. 202–3.

ing influence upon cultures: savagery, like civilization, was linked to certain climatic or geographical features. Nature, at least beyond the privileged shores of Europe, dictated culture.

It will be argued here that one of the principal manifestations of environmental otherness in European thought since the fifteenth century has been in terms of a developing distinction between temperate and tropical lands, and that the complex of ideas and attitudes that we will here call 'tropicality' represents environmentalism in one of its most influential and enduring forms. Part of the significance of 'tropicality' lay in its deep ambivalence. In part an alluring dream of opulence and exuberance – Edenic isles set in sparkling seas – the tropics also signified an alien world of cruelty and disease, oppression and slavery.

Tropical Edens

To begin to understand this kind of otherness we need to understand the tropics as a conceptual, and not just physical, space. What was it that informed the use of the term 'the tropics'? Of course, there is a large body of geographical, ecological and medical literature that finds no conceptual difficulty with the idea that the tropics actually exist. The region is located in the middle latitudes of the globe (between the tropics of Cancer and Capricorn, $23\frac{1}{2}$ degrees north and south respectively) though tropical or sub-tropical conditions are often seen to prevail over a considerably larger area. The tropics have their own climatic and ecological features, but can be further divided between the hot, wet tropics (such as the Amazon or Congo basins), the dry savannahs (such as those of eastern Africa) and alpine areas (like the high Andes). It was, however, the first of these that most typically represented the tropicality we are concerned with here.

Calling a part of the globe 'the tropics' (or by some equivalent term, such as the 'equatorial region' or 'torrid zone') became, over the centuries, a Western way of defining something culturally alien, as well as environmentally distinctive, from Europe (especially northern Europe) and other parts of the temperate zone. The tropics existed only in mental juxtaposition to something else – the perceived normality of the temperate lands. Tropicality was the experience of northern whites moving into an alien world – alien in climate, vegetation, people and disease. And this sense of the physical and cultural consequences of moving from one zone to another was more acutely felt in the Atlantic world, where the transition from temperate to 'torrid' was relatively rapid and where it was closely bound up with the Atlantic slave trade, than it was in the Indian Ocean or the Pacific.

The history of tropicality can be said to go back at least five hundred years to the first European voyages of discovery to Africa, Asia and the Americas, though it only developed fully in the eighteenth century with the increasing involvement of northern Europe (and latterly North America) in the equatorial regions. It may well be that a sense of tropical otherness was not very prominent in the minds of the first European travellers and adventurers, most of whom came from southern Europe and for whom the subtropical and equatorial regions of the globe did not appear strikingly different from the climate, vegetation and animal life they were familiar with in the Mediterranean and North Africa.

During his first voyage to the Caribbean, Columbus seems to have been in two minds – was this a familiar world or a strange one? In his journal he made repeated comparisons between the trees, birds and fish he saw there and those he knew in Spain: some seemed familiar, others wonderfully strange and suggestive of an earthly paradise. On 17 October 1492 he recorded in his log concerning a place he called 'Long Island':

> During that time I walked among the trees, which were the loveliest sight I had yet seen. They were green as those of Andalucia in May. But all these trees are as different from ours as day from night and so are the fruit and plants and stones, and everything else. It is true that some trees were of species that can be found in Castile, yet there was a great difference; but there are many other varieties which no one could say are like those of Castile or could compare with them.[2]

[2]   J. M. Cohen (ed.), *The Four Voyages of Christopher Columbus*, London, 1988, p. 66.

Two days later, after visiting 'Crooked Island', he recorded in similar uncertainty:

> Everything on all these coasts is so green and lovely that I do not know where to go first, and my eyes never weary of looking on this fine vegetation, which is so different from that of our own lands. I think that many trees and plants grow there which will be highly valued in Spain for dyes and medicinal spices. But I am sorry to say that I do not recognize them. When I reached this cape [which he called Hermoso, beautiful], the scent of flowers and trees blew offshore and this was the most delightful thing in the world.[3]

Had Columbus, seeking Cathay, stumbled on Eden? 'The singing of small birds is so sweet,' he noted on 20 October, 'that no one could ever wish to leave this place.' Such was his fascination with the environment that Columbus at first barely noticed the inhabitants at all, except in passing, as part of the landscape, between the birds and the trees. Until relations between the Spanish and the Arawaks turned sour and plunged into violence, the people Columbus encountered seemed to suit the innocence of the place, their near-nakedness a sure sign of their closeness to nature.

If, in the Americas, the first encounters of white explorers with the tropics invoked images of Eden, so too was Asia associated with the abundance of nature in the tropics and the kinds of riches Europe had hitherto only dreamed of. In 1527 Robert Thorne, an Englishman in Seville, reported that he had heard that the East Indies were 'fertile of cloves, nutmegs, mace, and cinnamon', and that the islands also abounded in 'gold, rubies, diamonds, jacynths, and other stones and pearls'. He added significantly, as were 'all other lands, that are under or near the equinoctial'.[4]

Visitors to south and south-east Asia marvelled at the luxuriant

---

[3] Ibid., pp. 68–9.
[4] Richard Hakluyt, *Voyages and Discoveries: the Principal Navigations, Voyages, Traffiques and Discoveries of the English Nation*, abridged edn, Harmondsworth, 1972, p. 49.

vegetation, the amazing colours and diversity of tropical birds and animals, and the sweet fragrance of flowers and spices that greeted them as they neared the shores of Malabar, Sumatra and the Moluccas. One sixteenth-century Portuguese writer described the island of Ceylon (Sri Lanka), with its palm groves, cinnamon trees, pure water and sweet oranges as being 'as if nature had made of it a watered orchard'; another pictured Bengal, with its abundant crops, livestock and game, as a land that was 'large, fruitful, and healthy'.[5] To such observers the tropics were, by contrast with a Europe scarcely free of famine and plague, a world in which nature was 'bountiful' and the climate 'healthy'.

Time – and closer acquaintance – served to strengthen at least some of these impressions of the benign and bountiful tropics, and the spread of printing gave such ideas a wide currency in Europe. As Philip Curtin observed a 'full-fledged myth of tropical exuberance' had come into existence in Europe by the eighteenth century, adding significantly that for Europe at the time 'the world between the tropics was much more of a "New World" than North America had ever been.'[6] Certainly, European impressions of the hot, wet tropics inclined much more to the paradisical than the pestilential. This perception of paradise was further strengthened by European encounters with Pacific islands, especially Tahiti, in the second half of the eighteenth century. Naturalists as well as novelists vied in their enthusiasm for the tropics as a source of pleasure and a welcome escape from cold and careworn Europe. Curtin quotes Henry Smeathman who visited West Africa on a botanical expedition and later recalled

Pleasant scenes of vernal beauty, a tropical luxuriance, where fruits and flowers lavish their fragrance together in the same bough! There nature animates every embryo of life; and reigning in vegetable or animal perfection, perpetually glows in wild splendour and unculti-

[5] Donald F. Lach, *India in the Eyes of Europe: the Sixteenth Century*, Chicago, 1968, pp. 343, 416.
[6] Philip D. Curtin, *The Image of Africa: British Ideas and Action, 1780–1850*, Madison, 1964, pp. 58–60.

vated maturity! ... I contemplate the years I passed in that terrestrial Elysium, as the happiest of my life.[7]

But an affirmative tropicality, combining a taste for the exotic, romantic and scientific, was most influentially articulated by the German traveller and naturalist Alexander von Humboldt, following his travels in Central and South America between 1799 and 1804. Humboldt surely did more than any other individual to 'invent' the tropics as a field of scientific enquiry and an aesthetic domain and to feed a positive perception of the tropics into wider environmental debates. His experience of the tropics left him with almost overwhelming impressions of 'organic richness' and 'abundant fertility'. 'Nature in these climes,' he wrote, 'appears more active, more fruitful, we may even say more prodigal of life.' The fecundity and diversity of the tropics fed his thoughts of how 'one sole and indissoluble chain binds together all nature' into a single, 'harmoniously ordered whole', which he called the Cosmos.

Driven as much by a Romantic appetite for sensation and sentiment as by an Enlightenment spirit of objective enquiry, Humboldt found in tropical America, with its snow-capped mountains, dense rain-forests and vast plains, a 'grand and imposing spectacle' that excited in him 'feelings of astonishment and awe'.[8] The tropics, he enthused, was a privileged site from which to observe and experience nature: 'Nowhere does she more deeply impress us with a sense of her greatness, nowhere does she speak to us more forcibly.' But, as Mary Louise Pratt has recently argued, Humboldt (for all his interest in Mexico's pre-Hispanic civilization) was far more inspired by the landscape than its inhabitants, who appeared dwarfed by the power and majesty of its mountains, rivers and plains. If, as she suggests, Humboldt

---

[7]   Ibid., p. 59.
[8]   Katherine Emma Manthorne, *Tropical Renaissance: North American Artists Exploring Latin America, 1839–1879*, Washington, 1989, p. 13; Alexander von Humboldt, *Cosmos: a Sketch of a Physical Description of the Universe*, vol. I, London, 1901, pp. 3, 5, 13–14.

was responsible for 'reinventing' America, it was America as nature, not as an anciently peopled place.[9]

Through his voluminous writings, Humboldt inspired countless naturalists, artists and travellers to see the tropics with fresh and wondering eyes. Attaching as much importance to the 'poetic imagination of mankind' as to nature's objective reality, he relished Columbus's descriptions of the Caribbean, and delighted in pictorial representations of tropical scenes. But he claimed that 'an inexhaustible treasure' still remained in the tropics, as yet 'unopened by the landscape painter',[10] and in both Europe and North America Humboldt's writings stirred unprecedented excitement in tropical America and its scenery. In the 1850s and 1860s a series of North American landscape painters undertook an artistic pilgrimage to the region to try to put on canvas what Humboldt had so inspiringly described in words. Among them was Frederic Church, a former pupil of Thomas Cole of the Hudson Valley School of painters in New England. His vast canvas *The Heart of the Andes*, exhibited in New York in 1859 (coincidentally the year of Humboldt's death), marked a high point of North American artistic interest in the tropics. Painting tropical landscapes was an attempt to capture the 'romance of the tropics' and Humboldt's awe at witnessing the vast and momentous forces of nature. But such paintings have also been understood (in the age of the Monroe Doctrine and Manifest Destiny) as an expression of pride in, and even proprietorship over, the entire American continent. And, significantly, people hardly figure in these grand, but empty, landscapes. Here was nature barely untouched by the 'transforming hand of man'.

The descriptions of the naturalists who accompanied Cook on his Pacific voyages of the 1760s and 1770s, along with Humboldt's scientific and travel accounts of Latin America, have been taken to mark the point at which 'the great world of the tropics' began to exert an exceptional influence upon European concep-

[9] Mary Louise Pratt, *Imperial Eyes: Travel Writing and Transculturation*, London, 1992, pp. 119–30.
[10] Humboldt, *Cosmos*, vol. II, pp. 370–2, 449.

tions of nature. The tropics seemed to bring scientists, as well as artists, closer to the secrets of nature. If not Eden, the tropics might yet yield insights into the mysteries of creation. It was arguably on Charles Darwin that Humboldt and his uniquely Romantic–scientific vision of the tropics had their greatest impact. Darwin later claimed that his 'whole course of life' was 'due to having read and re-read' Humboldt's *Personal Narrative* in his youth. When he first visited Brazil in 1833 during the voyage of the *Beagle*, he remarked with Humboldtian fervour: 'Here I first saw a tropical forest in all its sublime grandeur – nothing but the reality can give any idea how wonderful, how magnificent it is ... I formerly admired Humboldt, I now almost adore him; he alone gives any notion of the feelings ... on first entering the tropics.'[11]

Calling at Bahia on his return journey in 1836, Darwin noted how in the tropics, even in the vicinity of the towns, 'the wild luxuriance of nature ... overpowers in picturesque effect the artificial labour of man.' But he still found it hard to convey his emotion in words.

> Epithet after epithet is found too weak to convey to those, who have not visited the intertropical regions, the sensation of delight which the mind experiences ... The land is one great wild, untidy, luxuriant hothouse, which nature made for her menagerie, but man has taken possession of it, and has studded it with gay houses and formal gardens. How great would be the desire in every admirer of nature to behold, if such were possible, another planet; yet to every one in Europe, it may truly be said, that at the distance of a few degrees from his native soil, the glories of another world are open to him.[12]

The naturalists' response to the tropics was clearly of more than aesthetic importance. Biology and ecology were profoundly affected by this nineteenth-century encounter with the tropics,

[11]   Donald Worster, *Nature's Economy: a History of Ecological Ideas*, Cambridge, 1985, pp. 132, 137.
[12]   Charles Darwin, *Voyage of the Beagle*, Harmondsworth, 1989, pp. 366–8.

though it is striking that, where Humboldt found in the tropics evidence for an essentially harmonious Cosmos, Darwin discovered proof of a fiercely Malthusian struggle for survival.

The 'Torrid Zone'

It is clear, then, that for many Europeans from the fifteenth century onwards the tropics, and especially what Richard Grove has described as 'tropical island Edens',[13] constituted an attractive expression of exotic otherness. They were an escape from a bleak, oppressive, excessively humanized environment. They suggested a world where nature was, as it had been in the Garden of Eden, beautiful and bountiful, a green Utopia where people lived in harmony with their environment. None the less, Grove has argued, the fragility of these tropical island eco-systems was also increasingly apparent to visiting Europeans, especially in such places as St Helena and Mauritius. The destruction and desiccation caused by European intervention – through felling trees, hunting birds and animals to extinction, and creating plantations – created an awareness in the minds of Dutch, French and latterly British administrators and naturalists of the extreme dangers of rapid ecological change. Grove argues that an environmental awareness and early belief in the need for conservation measures thus developed in such oceanic outposts during the seventeenth and eighteenth centuries, long before George Perkins Marsh expressed concern about environmental degradation in America in the 1860s. Thus, according to Grove, 'the seeds of modern conservationism developed as an integral part of the European encounter with the tropics.'[14]

But it would be wrong to imagine (as Grove seems to do) that there existed a single, Edenic image of the tropics and that

[13] Richard H. Grove, *Green Imperialism: Colonial Expansion, Tropical Island Edens and the Origins of Environmentalism, 1600–1860*, Cambridge, 1995.
[14] Ibid., pp. 3, 240, 277, 475, 483.

environmentalism in the modern sense was the prevailing response, even among naturalists. On the contrary, alongside the awe and the intimations of an earthly paradise, there lurked an abiding sense of danger, alienation and repugnance, feelings more effectively expressed in Joseph Conrad's novel *Heart of Darkness* than in Bougainville or Cook's descriptions of Tahiti. Conrad was writing a century and a half after those Pacific pioneers, at a time when the interior of Central Africa was passing under European rule but still remained, in Western minds, a dark and disturbing presence. The long journey up the Congo River to meet the mysterious Kurtz (as Conrad describes it) presents a haunting, menacing contrast to earlier accounts of Tahiti and its inhabitants. The difference is not just between island and interior, or between the Pacific and Africa (for Africa, too, had its Edenic associations). Rather it was a difference between two contrasting understandings of the tropical world.

To this dark side of tropicality there was a history almost as long and as complex as the Edenic. Again, nature set the tone. Sir Walter Raleigh, who from his adventures in Guyana spoke with some authority on the matter, remarked in 1614 that 'if there be any place upon earth of that nature, beautie, and delight, that Paradise had, the same must be found within ... the Tropicks.' But he acknowledged that the tropics also had another, less attractive side, for they held 'the fearfull and dangerous thunders and lightnings, the horrible and frequent earthquakes, the dangerous diseases, the multitude of venomous beasts and wormes'.[15]

From the mid-eighteenth century onwards, negative representations of the tropics were becoming more commonplace in travellers' tales and even in fiction. Grove cites *Robinson Crusoe* as an illustration of the literary fascination with tropical islands (though it is hard to see Crusoe's life on his island as anything but hazardous). It should not be forgotten, however, that long before Daniel Defoe's hero was providentially shipwrecked off the

---

[15]   James Prest, *The Garden of Eden: the Botanic Garden and the Re-creation of Paradise*, New Haven, 1981, p. 34.

Brazilian coast, he had been a 'Guiney trader'. West Africa held out the prospect of riches, but it also meant the risk of being enslaved or attacked and eaten by the 'prodigious number of tygers, lyons, leopards, and other furious creatures which harbour there'. Violent images of the tropics also abounded in the medical and topographical literature that was beginning to spring up around European trading posts along the West African coast and in the Caribbean. Writers of the period frequently warned of the dangers of an intemperate climate and all manner of gnawing, roaring and biting creatures. In Africa, wrote James Lind, a British naval surgeon in 1768,

> all nature seemed at enmity with man. [Men] were prevented from walking in the woods by tygers . . . and if . . . they should escape these, they exposed themselves to the views of venomous serpents . . . The river swarmed with crocodiles: the earth had its white ants, the air its wild bees, its sand-flies, its mosquittoes.[16]

In an environment so inimical to human comfort, it was not surprising that diseases raged 'with peculiar violence' in the tropics. Some of these resembled the familiar fevers of Europe, but, as Dr John Hunter wrote of Jamaica in 1788, those in the tropics were 'greatly more violent in their attack, quicker in their progress, and more fatal in their termination' than was the case in Europe.[17] There is an important chronology here. Early works on the natural history and diseases of the West Indies, such as Thomas Trapham in 1679 and Hans Sloane some thirty years later, saw nothing exceptionally unhealthy or threatening about the West Indies. But by the time William Hillary wrote his medical *Observations on Barbadoes* in 1759, a stark contrast was being made between temperate and tropical climate and disease. Over the next fifty years the contrast was further developed in a great outpouring of medical treatises. The term 'tropical' became

[16]   James Lind, *An Essay on the Diseases Incidental to Europeans in Hot Climates*, London, 1768, p. 44.
[17]   John Hunter, *Observations on the Diseases of the Army in Jamaica*, London, 1788, pp. 14–15.

increasingly employed to express this negative otherness, as for instance in Benjamin Moseley's *Treatise on Tropical Diseases and on the Climate of the West Indies* in 1787, in which reference was made to the transition from temperate to hot climates as being 'so annoying to human nature' and in itself a predisposing cause of ill health. Moreover, these medical texts paid increasing attention to the health of African slaves, both as a subject of scientific curiosity and a source of danger to Europeans. The perils of the tropics were coming to be identified with race as well as place.

Behind this shift of emphasis from paradisical to pestilential there was, of course, a process of real change taking place with the mass importation of African slaves as a result of the sugar revolution and the corresponding transformation of the West Indies disease ecology. But the perception of the tropics was being fundamentally altered, with significant implications for other parts of the tropical world as well.

Although 'intemperance', 'venery', and inappropriate dress, diet and demeanour were thought to add to European vulnerability to disease, it was the tropical climate that was almost universally identified as the principal cause of deadly fevers and fluxes. The heat, humidity and rapid temperature changes of the tropics were believed to have a grievous effect on European constitutions (while 'native' peoples were presumed to be largely inured to such effects), and to predispose to disease, even where they did not actually cause it. The eighteenth-century revival of Hippocratic theories of environment and disease gave support to this negative representation, for the hot, wet tropics were thought to produce morbid miasmas on a scale, and of an intensity, unparalleled in Europe. Lind noted how, after the annual floods had retreated from the plains of Bengal, 'the air is . . . greatly polluted by the vapours from the slime and mud left by the Ganges, and by the corruption of dead fish and other animals.'[18] Small wonder, then, that Bengal was so notoriously prey to pestilential fevers. It also followed that doctors and surgeons, far from being the advocates of conservation that Grove describes,

[18]   Lind, *Essay on the Diseases*, p. 79.

often favoured the wholesale destruction of scrub and jungle in order to improve ventilation and dispel harmful miasmas. James Lind in the 1760s thought that Barbados had become much healthier after its tree cover had been removed to make way for sugar plantations. In the 1830s James Ranald Martin, a pioneer of medical topography in India, recommended the clearing of the entire forested tract of the Sunderbunds, 20,000 square miles in extent, to improve the health and climate of Calcutta (though there is no evidence that the Government of Bengal took his proposal seriously).

Despite sanitary amd medical advances in the nineteenth century, disease remained in the European mind one of the defining characteristics of the tropical world. The emergence of 'tropical medicine' as a medical specialism by the 1890s and 1900s served both to celebrate Europe's growing sense of mastery over the tropics and the abiding idea of tropical difference. The very idea of 'tropical diseases' and 'tropical medicine', always difficult to justify in purely epidemiological terms, since few diseases are in fact unique to the tropics, epitomized the way in which medical science in the imperial age gave its own endorsement to the idea of tropical otherness.

For Europeans, just living in the tropics was thought to be a physical and mental torment. In the early twentieth century, physicians identified what was called 'tropical neurasthenia', the state of acute anxiety induced merely by residence in the tropics. The sojourner, wrote Dr H. S. Stannus in 1926,

is exiled from home; often separated from his family . . . suffering in many cases loneliness and lack of congenial society . . . Living amidst a native population causes him annoyance at every turn, because he has never troubled to understand its language and its psychology. From early morn till dewy eve he is in a state of unrest – ants at breakfast, flies at lunch, and termites for dinner, with a new species of moth every evening in his coffee. Beset all day by a sodden heat, whence there is no escape, and the unceasing attentions of the voracious insect world, he is driven to bed by his lamp being extinguished by the hordes which fly by night, only to be kept awake by the reiterated cry of the brain-fever bird or the local chorus of frogs. Never at rest. Always an on-guardedness.[19]

It was widely believed that while Europeans might, at considerable risk to their health and sanity, find a temporary home in the tropics, they could never permanently settle there. Although a process of 'seasoning' and 'acclimatization' was acknowledged, it was generally held that Europeans in warm climates were eternally destined to remain 'exotics'. Lind explained that:

> Men who thus exchange their native for a distant climate, may be considered as affected in a manner somewhat analogous to that of plants, removed into a foreign soil, where the utmost care and attention are required to keep them in health, and inure them to their new situation; since thus transplanted, some change and alteration must happen in the constitutions of both. Some climates are healthy and salutary to European constitutions, as some soils are favourable to the production of European plants. But the countries beyond the limits of Europe which are chiefly frequented by Europeans, are very unhealthy, and the climate often proves fatal to them.[20]

Residence in the tropics was also thought to involve grave moral and racial hazards to whites. It was commonly believed that Europeans could not reproduce themselves in such adverse conditions beyond one or two generations, or, if they did so, it would be in a sickly and degenerate form. Equally it was believed that the physical laxity caused by tropical heat and humidity was matched by moral laxity. Where eighteenth-century travellers had seen sexual freedom as one of the more attractive attributes of society in the South Seas, late nineteenth- and early twentieth-century writers were extremely censorious and anxious about racial decay and 'miscegenation'. Ellen Semple, whose determinist views were noted in chapter 2, held in 1911 that 'Transfer to the Tropics tends to relax the mental and moral fibre, induces

---

[19] H. S. Stannus, 'Tropical neurasthenia', *Transactions of the Royal Society of Tropical Medicine and Hygiene*, 20, 1927, p. 330.
[20] Lind, *Essay on the Diseases*, pp. 2–3.

indolence, self-indulgence and various excesses which lower the physical tone . . . The presence of an inferior, more or less servile native population, relaxes both conscience and physical energy just when both need a tonic.'[21] Huntington warned in similar fashion that 'any young man of European race with red blood in his veins is in more danger of deteriorating in character and efficiency because of the women of the tropics than from any other single cause.'[22]

Not just morals – aesthetics, too, reflected a deep European ambivalence towards the tropics. Eighteenth- and nineteenth-century writers on the tropics drew almost habitually on the language of aesthetic appreciation developed by Edmund Burke in his *Inquiry into the Origin of our Ideas of the Sublime and Beautiful* and William Gilpin's notion of the 'picturesque'. In the passages quoted earlier from Darwin, for instance, both 'sublime' and 'picturesque' are used to describe the views and vegetation of Brazil. The islands of the West Indies and the coast of West Africa were often similarly commended for their natural beauty and 'wild exuberance'. But this appreciation of the 'sublime' and the 'picturesque', rather than contradicting the image of the perilous tropics, served, paradoxically, to give it even greater weight, for behind every enticing view and pleasing vista lurked a lethal miasma. The tropics were treacherous as well as dangerous, their beauty a deadly deception. Alexander Bryson, a British naval surgeon and medical statistician, made this point about the island of Fernando Po off the West African coast in the 1840s. 'The island, taken as a whole, is perhaps one of the most beautiful on the face of the earth,' he wrote enthusiastically; it had few rivals anywhere in its 'sublimity and grandeur'. And yet, as the medical statistics grimly confirmed, disease was rampant and there was not 'any spot in the whole known world more detrimental to health'.[23]

---

[21] Ellen Churchill Semple, *Influences of Geographic Environment on the Basis of Ratzel's System of Anthropo-Geography*, London, 1911, p. 626.

[22] Ellsworth Huntington, *Civilization and Climate*, 3rd edn, New Haven, 1924, p. 74.

[23] Alexander Bryson, *Report on the Climate and Principal Diseases of the African Station*, London, 1847, pp. 21–2.

During his four years of travels on the Amazon, Alfred Russel Wallace formed a very mixed impression of 'tropical nature'. He had set out for Brazil in 1848 with 'an earnest desire' to visit a tropical country, to see the 'luxuriance of animal and vegetable life said to exist there', and to witness 'all those wonders which I had so much delighted to read of in the narratives of travellers'. He certainly found much to excite his scientific interest – in the 'vigour of vegetation', the diversity of plants and trees, and the sheer vastness of the Amazonian forest, though, like many other writers of the period, it was the commercial potential of the forests that fired his imagination. While Wallace felt there to be 'grandeur and solemnity' in the Amazonian forests, he found little of the beauty he had expected.[24]

> The huge buttress trees, the fissured trunks, the extraordinary air-roots, the twisted and wrinkled climbers, and the elegant palms, are what strike the attention and fill the mind with admiration and surprise and awe. But all is gloomy and solemn, and one feels a relief on again seeing the blue sky, and feeling the scorching rays of the sun.[25]

He was homesick for the 'changing hues of autumn' and the 'tender green of spring'. The best of tropical vegetation, in his opinion, could not match the subtle beauty of an English spring – crab-apples in bloom, horse chestnuts and lilacs. There was, he thought, 'a greater mass of brilliant colouring and picturesque beauty, produced by plants in the temperate, than in the tropical regions'.[26]

Wallace was not alone in such opinions. Tropical forests might inspire admiration but they often failed to appeal to those more at home with temperate landscapes. Henry Marshall, a surgeon with British forces in Ceylon from 1808 to 1821, observed with

---

[24] Alfred Russel Wallace, *A Narrative of Travels on the Amazon and Rio Negro*, 1st published 1853, New York, 1969, pp. xi, 2, 305.
[25] Ibid., pp. 305–6.
[26] Ibid., pp. 307–8.

some regret that the interior of the island was almost entirely covered with forest.

> The scene is here extremely dull and uniform. There is nothing to diversify the prospect. One forest succeeds another, divided only by narrow strips of unwooded surface, which are for the most part uncultivated. The traveller is frequently in the flat country enveloped for many miles together in thick woods. In vain he expects, as he moves on, to meet with some variety, some circumstance, connected with the happiness or occupation of man, to interest his feelings . . . The silence of the forest is seldom interrupted, except by the cooing of wood-pigeons, or the rustling of wild animals, the only inhabitants of the deep jungles.[27]

A century later, the novelist Aldous Huxley struggled to find anything 'Wordsworthian' about nature in the tropics. The Central American jungle could, he admitted, be 'marvellous, fantastic, beautiful', but it was also 'terrifying' and 'profoundly sinister'. The mass of interwoven vegetation that had so enthralled Humboldt, Huxley found 'alien to the human spirit'. It was, he concluded, only possible to 'adore' nature in a country where it had been 'nearly or quite enslaved by man'.[28]

Tastes in tropicality have moved on. Today, we readily identify the tropics with vanishing rain-forests and tigers pushed to the edge of extinction, a whole world of nature imperilled by human greed and indifference. But it is salutary to recall how very differently, how negatively, the tropical environment was perceived and understood less than a century ago.

[27] Henry Marshall, *Notes on the Medical Topography of the Interior of Ceylon*, London, 1821, pp. 2–3.
[28] Aldous Huxley, 'Wordsworth in the tropics', quoted in David C. Miller, *Dark Eden: the Swamp in Nineteenth-century American Culture*, Cambridge, 1989, p. 141.

Civilization in the Tropics

The tropics were not just perceived to be the site of an exotic, luxuriant nature, on the one hand, or wild beasts and deadly fevers, on the other. The tropics were also identified with the people who inhabited the equatorial regions. Inevitably, this involved comparisons between the temperate and tropical zones and discussion of how far the non-European peoples who inhabited these regions were adversely fashioned by climate and disease. To a degree which may appear extraordinarily blind and bigoted today, there was until the 1950s an almost universal belief among Western writers that the tropics were as intrinsically unsuited to civilization as the temperate zones were ideally fitted for it. One basic assumption was that because nature in the tropics was so fecund, the few needs of 'native' peoples could be met with little mental and physical labour. Smeathman, whom we quoted earlier, remarked of West Africa in 1786:

> Such are the mildness and fertility of the climate and country that a man possessed of a change of cloathing, a wood axe, a hoe, and a pocket knife, may soon place himself in an easy and comfortable situation. All the cloathing wanted is what decency requires; and it is not necessary to turn up the earth more than two or three inches, with a light hoe, in order to cultivate any kind of grain.[29]

The peoples of the tropics were thus thought to live a life of ease. But was this a desirable state? Increasingly, Europeans saw such a life as mere indolence which lacked the necessary stimulus to work, accumulate wealth and build a civilization after the approved European model. The early envy of climates, vegetation and soils more generous than Europe's own turned to sour reproach and deep moral disdain. In a Europe proud of its own dedication to industry, it was impossible to regard tropical abundance as other than a distinct cultural disadvantage. Curtin quotes James Steuart in 1770:

[29] Quoted in Curtin, *Image of Africa*, p. 61.

If the soil be vastly rich, situated in a warm climate, and naturally watered, the productions of the earth will be almost spontaneous: this will make the inhabitants lazy. Laziness is the greatest of all obstacles to labour and industry. Manufactures will never flourish here . . . It is in climates less favoured by nature, and where the soil produces to those only who labour, and in proportion to the industry of every one, where we may expect to find great multitudes.[30]

Europeans in south-east Asia were similarly disdainful of the effects of environment on culture. Frank Swettenham in Malaya was one of many British commentators who wrote in this vein:

Less than one month's fitful exertion in twelve, a fish basket in the river or in a swamp, an hour with a casting net in the evening, would supply a man with food. A little more than this and he would have something to sell. Probably that accounts for the Malay's inherent laziness; that and a climate which inclines the body to ease and rest, the mind to dreary contemplation rather than to strenuous and persistent toil.[31]

Nor were such views of life and labour in the tropics confined to eighteenth-century moralists or nineteenth-century colonial administrators. Ellsworth Huntington was a fervent believer in the environmental inferiority of the tropics and their inability either to produce their own advanced societies or provide a congenial home for civilized whites. It was, he claimed, widely agreed that 'the native races within the tropics are dull in thought and slow in action'. Why?

[P]rogress in civilization demands that people raise a variety of food, and raw materials, and accumulate a surplus on which to live while making new discoveries and taking new steps in advance. It also demands that people be able easily to travel about so as to get

---

[30]   Ibid., pp. 61–2.
[31]   Frank Swettenham, *British Malaya*, quoted in Syed Hussein Alatas, *The Myth of the Lazy Native: a Study of the Image of the Malays, Filipinos and Javanese from the 16th to the 20th Centuries*, London, 1977, p. 43.

ideas and materials from other places. All this is more difficult in
tropical than temperate lands. Moreover the incentive to do so is
less than in cooler countries. Not only is there less need of clothing,
shelter and fire, but the absence of strong seasonal contrasts
removes one of the greatest of all stimuli to activity.[32]

Whatever the reasons offered, commentators had no doubt that
civilization was rare, if not impossible, in the tropics. If it existed
at all, it could only be as the result of migration and invasion
from more temperate zones, as (it was alleged) in the case of
Vietnam or India. How Mayan civilization could ever have
flourished in the forests of Central America was an abiding puzzle.

Although in identifying tropical otherness much emphasis was
placed on European discomfort and feelings of alienation, the
dependence upon non-white labour was also an important part of
how Westerners perceived and responded to the tropics. In
particular, the presence of slavery was central to ideas and images
of the tropics as they emerged between the sixteenth and nineteenth centuries. On the one hand, slavery was seen to be 'natural'
to these regions. With nature so bountiful, a surplus could only
be generated from people who were 'naturally lazy' and able to
meet their needs with minimal effort through some form of
coercion. Likewise, climate and disease made it impossible for
Europeans to provide the necessary labour themselves (despite
the evidence of the early history of white settlement in the West
Indies). Slavery was, then, not only a practical solution to the
labour problem, but symbolized the otherness of the tropics, the
lands where the ordinary laws of labour did not apply.

But, against those who argued that slavery was 'natural' and
necessary in the tropics, there were increasingly, by the late
eighteenth and early nineteenth centuries, those who regarded
slavery as a blight on the true nature of the tropics – its excitement
and exuberance. Nature, for Humboldt, was a 'free domain', and
it pained him to see Caribbean plantations 'watered with the
sweat of African slaves'. Rural life lost its charms when it was

[32]   Ellsworth Huntington, *The Human Habitat*, New York, 1927, pp. 96–7.

'inseparable from the sufferings of our species'.[33] Darwin, too, found slavery in Brazil distressingly at odds with the exhilaration tropical nature inspired in him. Maria Graham, an English woman who travelled to Brazil in the early 1820s, felt a similar conflict. At Pernambuco (Recife) in 1821, she was 'absolutely sickened' by the sight of a slave market, which sent her and her companions scurrying back to their ship with 'heart-ache' and a resolution that 'nothing in our power should be considered too little, or too great, that can tend to abolish or to alleviate slavery.' An amateur artist, she revelled in the sights of a tropical landscape while lamenting the plight of the slaves, and her frequent references to 'the evils of slavery' mingled incongruously with delight at the 'gay plants', 'still gayer' birds, and the multitude of 'sweet-smelling' tropical flowers'.[34] This juxtaposing of brutal slavery with natural splendour was an important and troubling part of outsiders' views of the tropical world.

There were, though, those who presented civilization in the tropics in far more affirmative terms. Just as Brazil, which retained its slavery into the 1880s, was often seen as representative of the most negative aspects of tropical life, so it could also be taken to stand for its most attractive and creative. Indeed, tropicality in Brazil became a kind of positive expression of an emerging national identity. The significantly named Tropicalista School of medical scientists in Bahia between the 1860s and 1890s sought ways of demonstrating that Brazil need not be as unhealthy or as backward as many Europeans assumed. Half a century later, the sociologist Gilberto Freyre hailed with enthusiasm Brazil's 'new world in the tropics'. He saw Brazilian civilization as a unique blending of Portuguese talent for exploration and assimilation with the peoples of the tropics and the distinctive physical qualities of the equatorial world. For him, Brazil was more than just a backward outpost of Europe. It was a new kind of

[33] Alexander von Humboldt, *Political Essay on the Kingdom of New Spain*, 1st published 1811, New York, 1972, p. 92.
[34] Maria Graham, *Journal of a Voyage of Brazil, and Residence There, during Part of the Years 1821, 1822, 1823*, 1st published 1824, New York, 1969, pp. 105–10, 116.

civilization that was emerging and adapting itself to 'conditions and possibilities' that were 'not European but tropical: tropical climate, tropical vegetation, tropical landscape, tropical light, tropical colours'. Freyre praised the work of the Brazilian scientists who had helped overcome two of the country's greatest afflictions, malaria and hookworm. 'These great Brazilian victories in the humanization of the tropics,' he observed, 'have contributed much to destroying the European idea that these evils are inseparable from tropical conditions.' He went so far as to suggest a new science, which he called 'tropicology', 'to deal with the adaptation of European science and technology to tropical situations'. This would be concerned with problems of health, agriculture, urbanization and regional planning in the tropics, but also with education, politics, psychology and 'mental hygiene', for 'the behaviour of man in the tropics has to be considered, in some of its aspects, in relation to situations and conditions peculiar to [the] tropical environment.'[35] But Freyre's was an exceptional voice: seldom have the virtues and potentialities of civilization in the tropics been so positively expressed.

Appropriating the Tropics

Thus far we have discussed tropicality as an idea and an image, but it is important to remember that the tropics were more than just a cultural construction, an exotic 'other' of European invention and imagining. In a very practical sense the tropics were also physically transformed under European tutelage on a scale and in a manner comparable to what happened to the temperate 'Neo-Europes' of the New World and Australasia. But they were not transformed into replicas of Europe: instead, they became complementary economies and ecologies, designed to serve needs and desires that the temperate lands could not satisfy. There were three main ways in which Europeans strove to incorporate and

[35] Gilberto Freyre, *New World in the Tropics: the Culture of Modern Brazil*, New York, 1959, pp. 7, 145–6.

subjugate the tropics: by controlling natural resources (especially vegetable products), by mobilizing non-white labour and by gaining mastery over 'tropical diseases'. Each of these interrelated forms of appropriation and control had important environmental consequences, and, as Freyre suggested, called for solutions that were often different from those adopted in Europe. Labour and disease we have already touched upon; let us now to turn to the third factor in the tropical equation, the exploitation of tropical plants.

While all three of these factors in the transformation of the tropics have close associations, the links between botany and medicine were particularly close. In medieval Europe, the study of plants was largely motivated by their medicinal use, as the illustrated herbals of the period show. At many medical schools in the seventeenth and eighteenth centuries, a study of botany and natural history was considered a necessary part of medical training, and doctors from Leiden and Edinburgh were often expert botanists as well. Many of the early studies of plants from the Americas and the East Indies were motivated by this practical, medicinal interest. A Europe ravaged by syphilis (the gift of the New World to the Old?), as well as by malaria, plague and other diseases, hoped to find in other continents a cure for its afflictions, while Europeans abroad believed that they had much to learn from local physicians in the treatment of unfamiliar diseases and conditions.

Although it is often assumed that Western science, technology and medicine grew autonomously out of Europe's own intellectual traditions and practical needs, it is increasingly clear how much their development was stimulated by European expansionism and informed by contact with other cultures. The growth of botany is one illustration of this. The first chair in botany was created at the University of Padua in 1533, and the establishment of botanic gardens for the systematic study and use of plants at such centres as Padua, Leiden, Leipzig, Paris, Uppsala and Edinburgh closely followed the first phase of European expansion and the importation of previously unknown plants from America and Asia. Botany strived to incorporate and systematize this vast and often bewildering knowledge, and could no longer rely on ancient texts or folk empiricism as a guide to the new-found riches of the plant world. The first 'physic gardens' were places where plants from around the world could be grown for their medicinal properties. Initially, with Europe still in its medieval dream-time, it was possible to think of these gardens as reuniting the scattered fragments of a lost Eden, bringing together God's plant creation for the first time since the Fall. But, by the eighteenth century, such ideas were fading as botanical gardens became pioneering centres of scientific research – at Uppsala under Linnaeus, at the Jardin du Roi in Paris under Georges Buffon and at London's Kew Gardens under Sir Joseph Banks and his successors.

Botanizing in Europe was matched and aided by parallel activities overseas. One of the first and most original investigations was the work of the Portuguese physician Garcia d'Orta in western India. In a book published after his death in 1563, d'Orta described some sixty medicinal plants and gave an extended account of local diseases and their treatment by Indian doctors. Significantly, d'Orta described many plants and drugs unknown to the Greek writers on whom Europe had previously relied so heavily. In a similar exercise early in the seventeenth century, the Dutchman Jacob Bondt (Bontius) gave a detailed account of the plants, animals and medicinal drugs of the East Indies. Nor was the New World forgotten: indeed, it constituted the largest single source of knowledge about new medicinal drugs. From the mid-sixteenth century onwards, texts appeared in Spanish, English and other European languages advocating the use of such American drugs as sassafras, tobacco, sarsaparilla and guaiac wood (the last two being especially recommended for the treatment of syphilis).

The practical importance of these botanical links between Europe, America and Asia was enormous, for they began the process of knowing the nature and properties of plants from other parts of the world and hence their exploitation for economic and medical purposes. Plants represent an example of the way in which Europeans not only took up for their own use selected items from other (including tropical) environments, but also absorbed elements of indigenous knowledge about those environments. Most drugs, foodstuffs and other vegetable products were not, in the first instance, 'discovered' by Europeans. A knowledge of their uses and properties had first been established by indigenous peoples and was only subsequently taken over and incorporated into European botany. One illustration of this was cinchona, the 'Peruvian bark' adopted from the Andean Indians by the Jesuits for the treatment of fevers and later the basis for the European treatment and prevention of malaria. What had once grown wild and provided the Indians with a basic febrifuge had become by the late nineteenth century a major plantation crop in the Dutch East Indies and British India and was reputedly one of the essential 'tools' of imperial expansion.[36]

Botanic gardens in Europe and overseas (one of the first was established by the Dutch at the Cape of Good Hope in 1654) served an expanding network of scientific knowledge and plant exchanges. From his base at Uppsala, Linnaeus used the Swedish East India Company's overseas trading operations and contacts to build up a worldwide collection of plants and send out expeditions to find new species. Sir Joseph Banks at Kew was similarly able to use an extensive network of correspondents, overseas traders, army, navy and East India Company surgeons to acquire plant specimens and seeds from the East Indies, China and the Americas. In this very practical respect, the growth of botanical knowledge was closely linked with Europe's overseas commerce and maritime power.

Significant, too, were the links being developed between different tropical territories in the Americas, Africa, Asia and the Pacific. These allowed the relocation of tropical plants between one territory and another in ways that directly served the economic and political interests of the imperial powers but also stressed the common identity and function of a tropical world increasingly characterized by sugar, coffee, cotton, rubber, cin-

---

[36] Daniel R. Headrick, *The Tools of Empire: Technology and European Imperialism in the Nineteenth Century*, New York, 1981, ch. 3.

chona and sisal. Of course, lateral exchanges of this kind had been taking place since the earliest phase of European expansionism. The Portuguese were instrumental in transferring a number of food crops, such as maize and cassava, from the Americas to Africa, India, south-east Asia and China. But, by the late eighteenth century, plant exchanges were being organized on a far more systematic and scientific basis as Europeans began to realize the economic rewards of transferring plants from one tropical location to another. In so doing they might hope to find a more effective way to feed slaves (the purpose of Captain Bligh's fateful voyage on the *Bounty* was to transport breadfruit seedlings from Tahiti to the West Indies), to protect state revenues and curb famine (one of the intended purposes of introducing sago palms and other exotic plants to India), or to evade the troublesome restrictions and taxes foreign governments placed on the export of such commodities as tea, rubber and cinchona by growing them in tropical possessions and making intensive use of slave or indentured labour. To the British, relative latecomers in the quest for tropical lands and resources, plant exchanges had a particular appeal. The wealth of colonies was seen to lie principally in tropical produce, but until the late eighteenth century many of the most productive areas (apart from the West Indies) lay in the hands of other European powers or outside European control altogether. The acquisition of India from the mid-eighteenth century onwards gave the British an opening into the tropical world, and botanical knowledge was extensively used to develop these new acquisitions to benefit the British economy.

However, the relationship between science and empire remains contentious. Some historians have seen the relationship between botany and overseas empire as a direct and instrumental one, scientific knowledge translating readily into economic gain and political power. In her study of the imperial role of the botanic gardens at Kew, Lucile H. Brockway argued in this vein that there is 'no way to draw the line between science, commerce and imperialism' in the work of those who collected seeds and plants for Kew or assisted in their transfer from one part of the tropical world to another. The new plantation crops thus established 'complemented Britain's home industries to form a comprehensive system of energy extraction and commodity exchange which for a time, in the nineteenth and early twentieth centuries, made Britain the world's superpower'.[37] Brockway supports this claim with studies of three of the most economically valuable plants involved: rubber, sisal and cinchona. Of the three, cinchona is perhaps the most dramatic as seeds were smuggled out of Peru in the mid-nineteenth century and shipped via Kew to India, where cinchona plantations were established to provide a reliable source of the anti-malarial drug quinine, used to protect European soldiers in India and Africa. Indeed, she argues, 'the colonial penetration of Africa in the late nineteenth century by the European powers was accomplished only after a cheap and reliable source of quinine was available . . . In sum [she concludes] quinine was an essential arm of mature British imperialism.'[38]

Brockway possibly overstates the importance of quinine to European imperialism in the late nineteenth century (and the role of malaria in inhibiting previous European expansion in Africa), but her general line of argument about botany's association with empire is clear and largely convincing. But many historians are indignant at the suggestion that science can be assigned such a narrowly instrumentalist role as the knowing and willing errand-boy of empire. They have sought to detach science from imperialism's blameworthy past and to stress the role of scientists as disinterested believers in progress and even enlightened reformers. Richard Grove, for example, has criticized writers like Brockway for attributing to scientists 'exclusively utilitarian and/or exploitative' motives. The ideas of naturalists and other scientists, he argues, often ran counter to the existing commercial and revenue policies of colonial governments. They were motivated by high scientific ideals and humanitarian concerns that often made them actual or implicit critics of empire rather than its abject servants.[39]

---

[37] Lucile H. Brockway, *Science and Colonial Expansion: the Role of the British Royal Botanic Gardens*, New York, 1979, pp. 6, 84.
[38] Ibid., pp. 7, 132.
[39] Grove, *Green Imperialism*, p. 8.

It is, no doubt, simplistic to assume that imperial science was a crude monolith, devoid of dissidents and motivated only by a desire to serve the immediate economic and political needs of the imperial power. But, equally, it is all too easy to exaggerate the degree of autonomy scientists enjoyed or to attribute to them present-day values and thereby ignore the almost overwhelming power of the imperial ethos. As is often the case, we have yet to arrive at a more balanced assessment of the role of science, including botany, in European expansionism and empire. But what surely has become clear is the extent to which Europeans sought, certainly from the eighteenth century and to some degree even earlier, to manipulate and control tropical environments and the plants, animals and peoples identified with them. Just as Europe was remaking the temperate regions of the globe, so was it refashioning the tropics to meet European wants and desires. But not every part of the non-European world was treated alike in either material or ideological terms, and it is worth considering through a brief case study of India, how one particular region was affected by Europe's growing environmental ascendancy.

9
Colonizing Nature

India and the 'Dispositions of Nature'

In many ways India under British rule was relatively resistant to the wider currents of tropicality described in chapter 8. Geography played a part in this. Much of the sub-continent lies north of the Tropic of Cancer: rather more of India than Brazil lies outside the tropics altogether. And while much of Bengal in the north-east and Kerala in the south-west (like neighbouring Burma and Sri Lanka) might appear to fit the model of the hot, wet tropics, the open plains of Hindustan and the Deccan seemed rather incongruous.

A more decisive reason for India's elusive tropicality was surely historical rather than geographical. In their encounters with India since the late fifteenth century, Europeans had been forced to recognize that they were dealing with powerful states and resilient cultures, not just with the raw forces of nature. The scholarly investigations of Sir William Jones and his fellow Orientalists in the late eighteenth and early nineteenth centuries revealed that India had an ancient civilization comparable in many respects to those of Greece and Rome. India, like Humboldt's equatorial America, had snow-capped mountains, vast plains and dense forests, but, to European eyes, they were not its principal attributes. Seldom, if ever, was India represented as some kind of tropical Eden. In writings about India the 'picturesque' was commonly invoked not to describe scenes of untouched nature, but Indian costumes, religious practices, colourful processions, busy street scenes, ancient temples and half-ruined palaces. It was perhaps only in the writings of medical men, especially those who had previously lived and worked in the West Indies or were influenced by that very different environmental and cultural encounter, that the word 'tropical' was consistently applied to India before the second half of the nineteenth century.

Europeans in India were all too conscious that India was not a 'virgin' or 'empty' land. Only in the remotest parts of the country – the high Himalayas and the densest forests of Kerala, Central India or Assam – was it possible to think of India as a 'wilderness', as when two British engineers found themselves deep in the Western Ghats in the early years of the nineteenth century and recorded:

> The larger portion of this desolate region is covered with lofty and luxuriant green forests in every direction . . . The whole scene is truly sublime; a large portion of this wild tract has not been explored for the want of guides, and the difficulty of penetrating such wild extensive regions.[1]

Over much of India, however, the environment had already been physically and culturally appropriated for thousands of years and in ways which could not be ignored even by the most ardent advocates of Europeanization. India's population had not been swept away by European diseases (indeed, India was rife with precisely the kinds of Old World diseases that had proved so destructive in the New World), and the landscape was encrusted with cultural markers of every description. The dusty plains around Delhi were thick with the forts, tombs, palaces and

[1] Benjamin S. Ward and P. E. Connor, *Geographical and Statistical Memoir of the Survey of the Travancore and Cochin States*, Travancore, 1863, pp. 205–6, quoted in Richard P. Tucker, 'The depletion of India's forests under British imperialism: planters, foresters, and peasants in Assam and Kerala', in Donald Worster (ed.), *The Ends of the Earth: Perspectives on Modern Environmental History*, Cambridge, 1988, p. 120.

gardens of the Mughals and their Muslim, Hindu and Buddhist predecessors. And so it was over much of India. Every idyllic spot for a European picnic had been pre-empted by a Hindu shrine or was guarded by the tomb of a Sufi saint. At the foot of each shady tree lay a vermillion-stained stone representing a folk deity, offering protection against smallpox or cholera. The cows that roamed the streets, the peacocks that strutted in the parks, the tigers that roared in the jungle – few aspects of the Indian environment, animate or inanimate, seemed free and unappropriated by some aspect of Indian, especially Hindu, culture.

However, by the late nineteenth century, India's incorporation into the tropics was becoming increasingly evident. Its diseases, vegetation, soils, climate, agriculture, even its people, were steadily being brought within the framework of tropicality. One can see this as, in part, a consequence of the growing authority of imperial science and the connections being made between the several parts of the tropical empire. Whether it could boast of an ancient civilization or not, India, like many another tropical estate, grew tea and coffee, cotton, rubber and cinchona, and Indian indentured labourers replaced African slaves as the primary source of migrant labour in tropical plantations from Fiji to Guyana. In the minds, too, of geographers like Semple and Huntington India displayed many of the characteristic features of a tropical country: an energy-sapping climate, debilitating diseases, an environment intrinsically unsuited to civilized life as they understood it. As imperial power grew, and with it Britain's sense of racial and technological superiority over India, so the environment was invoked more and more to explain the great gulf that divided them.

For instance, much importance was attached to the role of India's monsoons. The annual cycle of south-west and north-east monsoon rains was not only said to be the 'primary fact' in the meteorology of India, but also in many aspects of its economic and social life, and hence was an index of the power which 'the dispositions of nature' exercised over the Indian people. The monsoons provided the rain needed to sustain agriculture across a large part of the sub-continent, but if, as happened all too often in the second half of the nineteenth century, the rains failed, then drought and famine almost invariably followed. The monsoon was, according to the imperial geographer T. H. Holdich, 'the yearly salvation' of the millions of Indians who lived on 'the fruits of the soil. A good or bad monsoon is the criterion of plenty or of famine. By it India lives; without it there is starvation, death, and misery.'[2]

British colonial officials in nineteenth-century India perhaps found this correlation between monsoon failure and mass famine particularly significant because Britain had itself only recently emerged from a similar subjection to nature, but now prided itself that such catastrophic events could only happen in more backward, less enlightened countries. In Britain (with the exception always of Ireland) nature had been tamed.[3] But in India its influence lived on and affected almost every area of society and government. It was noted, for instance, how food riots and property crime burgeoned with the failure of the monsoon, as prices soared and food grains fell into short supply. The failure of the rains filled the prisons with 'hungry people not ordinarily criminal'. It drove desperate men to murder and robbery, women to petty theft and suicide. It also, more positively from the imperial perspective, provided the incentive for poor peasants and landless labourers to leave their villages and migrate to the tea gardens of Assam or the cane-fields of Natal.

Even when the monsoon did not fail, its effect on the health of the masses could still apparently be enormous. Public health reports and statistics showed how the monsoons brought in their wake epidemics of cholera, dysentery and other intestinal diseases. This correlation between climate and disease prompted one medical officer, writing in the *Imperial Gazetteer of India* in 1909, to draw a more general environmental contrast between India and Europe:

---

[2] T. H. Holdich, *India*, London, 1904, p. 348.
[3] For an expansion of this argument, see David Arnold, *Famine: Social Crisis and Historical Change*, Oxford, 1988, esp. ch. 6.

The general state of the public health in every country depends on the measure of adjustment of the relations of the individual and the race to the environment: the more complete and continuous the adjustment, the greater the longevity. The tendency of European civilization is to give man more and more complete control of his surroundings, whereas in India these are actually and relatively stronger, more capricious and unreliable than in the West, while the individual is less resistant and adaptable. These influences have moulded the moral and physical character of the people and their civilization.[4]

The articulation of such extreme determinist views creates problems for the latter-day historian. Should these opinions be dismissed in their entirety as the product of an outmoded, self-justifying imperialism, or should it be recognized, in the context of a new social history of India, that such major environmental events and calamities as the onset or absence of the monsoon did indeed profoundly affect the lives and livelihoods of the great majority of the Indian population? It ought, in theory, to be possible to separate the imperial rhetoric from the analysis of material reality. In practice, however, the distinction is often hard to maintain for the very sources themselves are imbued with a particular mind-set, a colonial way of understanding nature and representing its human consequences.

In India, moreover, there was a recognition that in a country of almost continental proportions there was bound to be a great diversity of climatic and other environmental factors at work on human minds and bodies. Hence determinist ideas were often used not just to try to encompass and characterize India as a whole but also to differentiate between its different regions and peoples. In an essay on 'Population' in the same volume of the *Imperial Gazetteer*, E. A. Gait linked ideas of racial origins and identity with aspects of India's physical environment to explain significant differences between its peoples. His argument reflected the idea, as old as Hippocrates, that harsh environments bred tough, resourceful individuals, while warm and fruitful lands

---

[4]   *The Imperial Gazetteer of India*, vol. I, London, 1909, p. 500.

fostered indolence. But to this old formula was now added the fashionable new ingredient of Darwinian struggle:

> In the north-west the dry climate, and the incessant struggles with man and nature in which only the fittest could survive, have combined to produce a brave and hardy race of good physique; while the easy life in the steamy and fertile rice plains of the Gangetic delta, though encouraging a rapid increase in the number of its inhabitants, has sapped their energies and stunted their growth. The small, weak, and timid Hindu peasant of Bengal differs from the tall, sturdy, and brave Sikh, or the turbulent and active Pathan, to a greater degree than does the Scandinavian from the Spaniard or the Englishman from the Turk.[5]

From the imperial perspective, the value of such determinist and reductionist modes of reasoning was clear. India was subject to nature to a far greater degree than Europe: hence its backwardness, inferiority and internal divisions; hence, too, the need for the British to rule over India, to bring about 'improvement', 'order' and 'progress', and so liberate Indians from their servitude to nature. However, at a time when nationalist ideas were gaining strength in India, nature could also be invoked to show that India was not one nation, and would not become one in the forseeable future. Nor could it even lay claim to be a civilization, whatever its ancient achievements may have been. Environmental forces – climate and disease above all – were repeatedly invoked to demonstrate and explain Indians' moral and physical weakness and to justify the indefinite continuance of imperial rule.

It was, accordingly, one of the self-declared tasks of colonial sanitation and medical science to rescue Indians from their pathetic thraldom to nature, while at the same time demonstrating their inability to do this for themselves. Ronald Ross of the Indian Medical Service, best known for his discovery in the 1890s of the role of mosquitoes in the transmission of malaria, observed in his memoirs that Europeans who visited India for the first time were often struck by the sickly character of the

---

[5] Ibid., p. 447.

people. They appeared 'hard-working . . . faithful, docile, and intelligent', and yet in many parts of the country they possessed 'amazingly delicate physiques combined with great timidity and a habit of unquestioning obedience'. What, Ross asked, was the cause of this weakness? Was it poverty, climate or disease? His answer was malaria which, over the ages, had made Indians an 'ancient race outworn'. Through his medical research, Ross sought to discover the cause of malaria, asking, in the course of one of his poems, 'In this, O Nature, yield I pray to me.' In 1895, when Ross made his final breakthrough, he broke into triumphant verse:

> This day designing God
> Hath put into my hand
> A wondrous thing. And God
> Be praised. At His command,
> I have found thy secret deeds
> Oh million-murdering Death.[6]

Ross's personal sense of mission can, without too much exaggeration, be translated into a wider sense of imperial purpose. Although India was a populous country, although it was acknowledged to have had an ancient civilization whose monuments everywhere littered the landscape, it had (in the imperial estimation) failed in one essential respect – to overcome the elementary forces of nature. Thus, it was in no small part through a transformation of the environment, by establishing and demonstrating mastery over nature itself, that the British sought to advance and legitimize their rule in India.

Transforming India

In recent years, some historians have begun to argue that British rule had profound consequences on the Indian environment. This claim can be seen as part of a wider debate about the British

[6] Ronald Ross, *Memoirs*, London, 1923, pp. 97, 226.

impact on India. Environmental change is cited as evidence of how great that impact was, even in rural areas remote from the main urban centres of British power and influence, in contrast to those who have seen colonial rule as having had only a relatively superficial influence on India, especially in the countryside. But this impact is not seen in the heroic manner that Ross described. Instead, in a significant inversion of colonial ideals and self-congratulatory rhetoric, it is the negative consequences of British rule which are now more commonly stressed through the environmental debate rather than its once much-trumpeted 'achievements'.

At the same time one can see this reappraisal of India's environmental history as a critical riposte in the debate about 'ecological imperialism'. India clearly did not become a 'Neo-Europe' in the way that parts of the Americas and Australasia did, according to Crosby. It was not thrown open by disease and the invasion of new plants and animals to European settlement, nor was the indigenous population eliminated or even demographically marginalized. The arrival of Vasco da Gama and the Portuguese in south-western India in 1498 was not, ecologically speaking, an epoch-making event comparable to the impact of the Spanish in the New World from 1492 onwards. Certainly, India had some share in the trans-oceanic 'exchanges' of the early phase of European expansion. Syphilis reached India (probably via Europe) from the Americas, but, for all its horrors, this can have contributed little to the sum of Indian mortality. Several plants of American origin, including chilies, tomatoes and groundnuts found their way to India and helped give it its distinctive modern cuisine, but the spread of these new food crops, along with other imports like tobacco, took place within the context of an existing system of peasant agriculture and was relatively slow and undramatic. None the less, over a longer span of time, significant changes did occur. Without ever becoming a 'Neo-Europe', India's ecology was profoundly affected by increasing European contact and especially by colonial rule from the mid-eighteenth century onwards. Just how profound the process of environmental change actually was, historians are only now beginning to discover. But one historian has remarked that, by 1947 (the year the British departed),

> nearly all of the vegetation map of India was determined by the intricate structure of power which had evolved under the British raj. The lands of the subcontinent were almost entirely domesticated, under the most complex system of resource extraction which any European empire ever established in the developing world.[7]

The British contribution to Indian environmental change is a vast subject and a few leading examples must suffice. We have already touched upon drought and famine, and it is therefore appropriate to begin with water husbandry. India had an extensive system of rain- and stream-fed irrigation, by tanks (reservoirs), canals and dams, long before the advent of British rule. Some, but by no means all, of these systems had fallen into disuse by the mid-eighteenth century when the British, operating through the English East India Company, began to establish themselves as a major territorial power in south Asia. From the 1830s, British engineers began to re-open old irrigation works and create extensive new ones. One of the largest of these schemes, the Ganges Canal, was opened in 1854; others followed, tapping the waters of the Godavari, Jumna and Indus rivers. The need to check drought and famine was one incentive behind this expansion of irrigation, but so were more immediate revenue considerations, since irrigated land could produce two or three crops a year in place of only one, and often precisely those crops that had the highest market value. By the 1890s, nearly 44,000 miles of main canals and distributaries had been constructed in British India, irrigating more than 13 million acres. Fifty years later, towards the close of the British period, the figure had been increased to 75,000 miles and 33 million acres, a quarter of India's total cropped area. Many parts of India were physically transformed in the process, none more so than Punjab where the creation of an intensive river-fed irrigation system and the establishment of

---

[7] Tucker, 'The depletion of India's forests', p. 140.

purpose-built 'canal colonies' radically changed the appearance and use of the land. Previously largely arid, Punjab became, and remains, one of the sub-continent's most prosperous cereal-producing regions.[8]

To the British, irrigation works on such a vast scale and in the face of enormous technical difficulties were an outstanding monument to Western science and technology. They epitomized colonial faith in the mastery of nature and the material benefits that accrued from it; they were a visible and enduring statement of the power and resources of the colonial state. Moreover, in offering an apparent solution to the age-old problem of famine, they represented in British eyes, and (it was hoped) those of their Indian subjects, clear evidence of the benefits of firm but paternalistic rule. But they were not always, environmentally speaking, beneficial. The canals brought huge problems in their wake, including waterlogging and salination (the impregnation of the soil by salts brought to the surface by the seepage of water from canals and irrigated fields). In many parts of northern India the advance of the canals was matched by the spread of malignant strains of malaria as irrigation ditches and waterlogged soil provided new breeding places for anopheles mosquitoes. A deterioration in Indian health was one of the negative consequences of colonially managed environmental change.[9]

Railways provide a second example of the far-reaching character and consequences of India's environmental revolution. From modest beginnings in 1853, the country's railway system expanded rapidly after the Rebellion of 1857–8, spurred on by strategic as well as commercial considerations and attractive investment opportunities. By 1910, India had the fourth largest railway network in the world, with more than 32,000 miles of

[8]   Elizabeth Whitcombe, 'Irrigation', in Dharma Kumar (ed.), *The Cambridge Economic History of India*, vol. II, Cambridge, 1983, pp. 677–737.
[9]   Elizabeth Whitcombe, 'The environmental costs of irrigation in British India: waterlogging, salinity, malaria', in David Arnold and Ramachandra Guha (eds), *Nature, Culture, Imperialism: Essays on the Environmental History of South Asia*, Delhi, 1995, pp. 237–59.

track.[10] Although the economic impact of the railways has been the main focus of debate in the past, the environmental consequences now appear equally significant. Vast engineering works were needed to span the flood-plains of India's rivers, to traverse mountainous ghats or climb steep inclines to hill-stations like Simla and Darjeeling. In the process, a great deal of forest and other natural vegetation was cleared and the soil exposed to erosion. For purposes of construction – for bridges and sleepers – and also initially as fuel for the locomotives, vast quantities of timber were felled, transported and consumed. The railways opened up for commercial exploitation hitherto inaccessible forests: an accelerated process of deforestation was one of the principal legacies of the railway age. Like the irrigation canals, railways also had an adverse impact on the disease environment. The new embankments interfered with natural lines of drainage and so contributed to the waterlogged conditions in which malarial mosquitoes bred.[11] Greater human mobility by rail made it easier for diseases to move rapidly from one area to another: the spread of bubonic plague in the 1890s and influenza in 1918–19 are examples of this rapid dissemination by means of India's railways.

Colonial Forestry

Alongside the environmental impact of irrigation canals and railways can be set the fate of India's forests, now seen to have been a major factor in making the colonial era into what Madhav Gadgil and Ramachandra Guha have aptly called a 'watershed in the ecological history of India'.[12] As in North America, the

[10]   John M. Hurd, 'Railways', in Kumar (ed.), *The Cambridge Economic History of India*, vol. II, pp. 737–61.
[11]   Ira Klein, 'Malaria and mortality in Bengal, 1840–1921', *Indian Economic and Social History Review*, 9, 1972, pp. 132–60.
[12]   Madhav Gadgil and Ramachandra Guha, *This Fissured Land: an Ecological History of India*, Delhi, 1992, p. 116. However, there had been extensive deforestation and ecological change in some parts of India in the centuries immediately

depletion of India's once vast forests represents environmental change at its most profound and dramatic. It is now almost impossible to envisage the extent of forests in India as they were even in the eighteenth century – and of the diversity of human activity and the plants and animals associated with them.

As in Europe, the size of India's forests had fluctuated over the ages as a result of agrarian expansion and contraction and as wars, famines and epidemics took their toll. Before 1800, many writers maintain, there was a rough equilibrium between human needs and nature, and though the extent of this harmony can easily be exaggerated, difficulties of access and extraction, the limited commercial value of timber, the resistance of forest-dwelling peoples to outside control, and the malarial conditions that abounded in such areas as the Terai (along the present-day border between India and Nepal), and the protection afforded by sacred groves and the hunting rights of rajas and the imperial Mughals – all of these may well have contributed to the long-term preservation of large areas of India's forests. But the picture cannot, even so, have been quite as idyllic or harmonious as some writers would like to imagine.[13] The need for fuel and building materials for India's cities had for centuries made heavy demands on neighbouring forests, and in times of warfare even sacred groves might be wrecked by bands of armed warriors or to clear a path for advancing armies.

None the less, India's forests were undoubtedly subject to unprecedented change during the course of the nineteenth century. The greatest pressures came from the expansion of peasant agriculture, the creation of tea and coffee estates, and the development of commercial forestry under state auspices. The value of Indian timber, especially teak, sal and sandal-

---

before British rule; see Richard M. Eaton, *The Rise of Islam and the Bengal Frontier, 1204–1760*, Berkeley, 1993.

[13] For a warning note about the dangers of attributing a modern environmental consciousness to pre-colonial forest-dwellers, see David Hardiman, 'Power in the forests: the Dangs, 1820–1940', in David Arnold and David Hardiman (eds), *Subaltern Studies VIII: Essays in Honour of Ranajit Guha*, Delhi, 1995, pp. 89–147.

wood, increased as the market economy grew and as hitherto inaccessible stands of trees could be reached by road and rail. Timber resources came to be regarded by the British as among the most valuable prizes to be won through territorial acquisitions, including the Terai in 1816, the teak-rich Burmese province of Tenasserim in 1826 and the north Indian kingdom of Awadh in 1856. From the mid-nineteenth century, the railways' demand for wood was almost insatiable: a single mile of broad-gauge track was said to require up to two thousand sleepers, and by 1878 alone over two million had been used in railway construction. Expanding colonial cities the size of Calcutta, Bombay and Madras made heavy demands for fuel and construction materials from an ever-widening hinterland, and land was cleared in previously wooded upland areas like Assam in the north-east and the Nilgiris in the south for growing tea, coffee and other plantation products.

The British also assisted the advance of this ecological frontier through the spread of peasant agriculture into previously unculti- vated or thinly populated tracts like Lower Burma as a way of augmenting government revenue and curbing land pressure and famine in more populous areas. India's forests had long been the home of supposedly 'primitive' peoples, especially tribals or *adivasis* ('original inhabitants') whose way of life was seen by the nineteenth-century British as unproductive or too backward and subsistence-orientated to serve the wider economic needs of the colonial economy. Many of these forest peoples were brushed aside in the cause of 'progress' and 'improvement', designated members of 'criminal tribes', or obliged to swell the ranks of agrarian labourers. Forests were seen as an obstacle to settled government and productive agriculture in a further sense, too: they harboured bandits, Thugs, and, as the events of the Rebellion of 1857–8 in northern and central India demonstrated, insurgents and mutineers. Cutting down the forests was thus a way of advancing the frontier of effective administrative control and minimizing sites of lawlessness and resistance. 'An open landscape provides little refuge for rebel, fugitive, or bandit. Just as the colonial regime cleared the countryside of firearms following the

Mutiny, so also did economic development clear away the natural
defences of local society against state power.'[14]

Although colonial policies did much to harm the forests and to
promote their ruthless exploitation, it is worth noting that the
British saw themselves in the opposite light: as trying to protect a
valuable natural resource and to save the forests from a process of
continuing destruction. In the 1920s, one leading forestry official,
E. P. Stebbing, argued that a 'war against the forests' had been
going on in India for centuries (he estimated for '3,500 years or
more'). He believed that many areas of India, once densely
populated, had been reduced to uninhabitable jungle and scrub
by 'the reckless, continuous and wholesale burning of the forests'.
Over many generations this had resulted in 'the gradual decrease
of water in the larger rivers . . . the drying up of springs, small
streams and rivers and . . . a gradual decrease in the rainfall of
the country'. Like many colonial foresters, Stebbing blamed the
'pernicious system' of shifting cultivation practised by many of
India's forest peoples. This was, in his view, 'the most wasteful of
all methods of forest utilization'; it resulted in 'thousands of
square miles of valuable forests' being laid waste every year and
'formerly fine timber forests being replaced by worthless scrub'.[15]

It must be doubted whether shifting cultivation did as much
damage as the British claimed or that its effects were anything
like as great as those of commercial forestry or plantation agriculture. But, by the 1850s and 1860s, the Government of India had
grown sufficiently concerned about the rapid depletion of timber
resources, and the supposedly harmful consequences of shifting
cultivation and related practices, to feel the need for state inter-
vention on a large scale. In 1864 it set up the Indian Forest
Service, followed by two far-reaching Forest Acts in 1865 and
1878. The acts created reserved state forests, estimated by 1900
to cover a fifth of the entire land area of British India, and

[14]   J. F. Richards, James R. Hagen and Edward S. Haynes, 'Changing land use
in Bihar, Punjab and Haryana, 1850–1970', *Modern Asian Studies*, 19, 1985,
p. 725.
[15]   E. P. Stebbing, *The Forests of India*, vol. I, London, 1922, pp. 32, 35–6.

established one of the largest state-run forestry enterprises any-where in the world. By 1947, the year of Indian independence, 99,000 square miles of forests had been brought under state control. The acts resulted in an 'unprecedented degree' of inter-vention in the lives of India's villagers. In the name of forest conservation, grazing was prohibited and shifting cultivation was banned or severely restricted. A small army of subordinate forest officers was employed to enforce these regulations, often in the face of bitter opposition from the local population who saw their traditional rights and means of subsistence substantially curtailed. In acts of 'everyday resistance' peasants and *adivasis* defied the forest laws to gather fuel, collect manure or graze their animals. Some forest-dwellers fled to India's princely states in order to evade the new forest restrictions, while others participated in a series of major rebellions between the 1870s and 1920s, in protest against the loss of their customary rights and in order to free themselves of such irksome constraints.[16]

Why did the British intervene in this way? Were they motivated by a genuine interest in forest conservation or simply responding to more immediate commercial and revenue considerations? Gadgil and Guha have argued that the British were, initially at least, little interested in conservation for its own sake: 'the imperatives of colonial forestry,' they observe, 'were essentially commercial.'[17] State-managed forests became a significant source of income at a time when government revenues were hard-pressed. Moreover, many of the forests were not so much con-served as commercially managed or even transformed in the search for profit maximization by, for instance, replacing mixed forests with plantations of teak, sal or other valuable hardwoods. By contrast, from a position far more favourably disposed towards colonial science and its 'green' credentials, Richard Grove has maintained that British forestry officials in India were actually

---

[16] Madhav Gadgil and Ramachandra Guha, 'State forestry and social conflict in British India', in David Hardiman (ed.), *Peasant Resistance in India, 1858–1914*, Delhi, 1992, pp. 258–95 (first published in *Past and Present*, 123, 1989, pp. 141–77).

[17] Ibid., p. 261.

working against the government's short-term financial interests and from a genuine commitment to the principles of forest conservation. In his view 'ecological constraints . . . had an exceptional impact on what one is accustomed to think of as the economic priorities of the colonial state.'[18]

Reappropriating Nature

The far-reaching material and ideological changes introduced by the British, and epitomized by the rapid depletion or changing character of India's forests, did not go unchallenged. The environment became in India (as in many other colonial territories and as in Europe itself) a major site and source of conflict between the rulers and the ruled. As just indicated, one aspect of this contestation was the attempt by India's rural populations to defend their customary uses of the forests from colonial regulation, and forest grievances were a factor in a number of rural revolts from the 1870s onwards. Resistance also came in the form of forest *satyagrahas*, or Gandhian-style civil disobedience campaigns, during the 1920s and 1930s, directed against corrupt or tyrannical forest guards and officials and against restrictions on the collection of leaf manure and fuel and the grazing of sheep, goats and cattle. Similar forms of popular resistance, based upon both practical grievances and a very different cultural perspective from the British, occurred in other areas, too, for instance in opposition to colonial public health campaigns or restrictions on fishing and hunting rights.[19]

[18]   Richard H. Grove, 'Colonial conservation, ecological hegemony and popular resistance: towards a global synthesis', in John M. MacKenzie (ed.), *Imperialism and the Natural World*, Manchester, 1990, p. 18. See also Mahesh Rangarajan, 'Imperial agendas and India's forests: the early history of Indian forestry, 1800–1878', *Indian Economic and Social History Review*, 31, 1994, pp. 147–67.
[19]   For some examples of these differences and conflicts, see David Arnold, 'Smallpox and colonial medicine in nineteenth-century India', in David Arnold (ed.), *Imperial Medicine and Indigenous Societies*, Manchester, 1988, pp. 45–65; Peter Reeves, 'Inland waters and freshwater fisheries: some issues of control, access and conservation in colonial India', in Arnold and Guha (eds), *Nature, Culture, Imperialism*, pp. 260–92.

But these manifestations of popular resistance to British environ-mental intervention and control are by no means the whole story. There were also forms of ideological resistance, most evident among the Indian middle class, to colonial attempts to appropriate the environment exclusively to a British agenda of 'civilization' and 'progress'. Indians sought, at various levels, to contest these col-onial 'readings' of their environment, to reject the negative images and associations that Western 'tropicality' imposed upon them, and to assert their own identity through environmentally related symbols and values. Gandhi's use of the colonial salt tax as the central issue in his civil disobedience campaign of 1930 was emblematic of this wider determination to contest British environ-mental control and to exploit its symbolic as well as practical significance. If nations are understood in Benedict Anderson's sense as 'imagined communities' then (as we saw earlier in the case of the United States) their use of certain images and symbols, drawn from the environment, have often been a powerful emotional rallying point and a focus for an emerging sense of national identity.

With its enormous ecological as well as cultural diversity, India was hard put to find a single set of environmental sites and symbols for this purpose. Salt, as Gandhi was quick to ascertain, was somewhat exceptional in having an all-India significance. The landscapes, the flora and fauna of Bengal, even the kinds of food people ate there, were very different from those of Gandhi's Gujarat in the west or Tamilnad in the extreme south. Moreover, many natural symbols in India carried a specific religious connotation and so might not be readily acceptable to followers of other faiths. The potent imagery of the cow, for instance, was fully exploited to rally Hindus, but it tended to have a divisive effect on Hindu–Muslim relations. Nor was it easy to separate India's rivers and mountains from a distinctly Hindu sacred geography.

None the less, environmental imagery was important in devel-oping a national as well as regional or religious consciousness in India and in contesting British cultural and political dominance. For instance, pride in the landscape of Bengal (as in its language and culture) was a conspicuous feature of the work of the Bengali poet and novelist, Rabindranath Tagore. His loving appreciation of the landscape of his home region was a combination of many cultural ingredients (from the inspiration of the Sanskrit poet and playwright Kalidasa to European Romantic writers and painters). But, not surprisingly, considering how disparaging the British had become about tropical Bengal and its 'weak' and 'effeminate' inhabitants by the late nineteenth century, there was also a note of patriotic defiance, as when the young Tagore wrote in September 1894:

> Many people dismiss Bengal for being so flat, but for me the fields and rivers are sights that I love. With the falling of evening the deep vault of the sky brims with tranquillity like a goblet of lapis lazuli; while the immobility of afternoon reminds me of the border of a golden sari wrapped around the entire world. Where is there another land to fill the mind so?[20]

One also finds in Tagore the reverse image of this patriotic romanticism – a nationalistic invocation of the environment as a reflection of India's depressed and degraded state under foreign rule. In a letter written only two days before the passage just cited, Tagore wrote in distress and despair at the conditions he observed in some of the poorer villages of riverine Bengal.

> Every home has rheumatism, swollen legs, colds or fevers, or a malaria-ridden child ceaselessly crying whom no-one can save. How can men tolerate for one minute such unhappy, unhealthy, unlovely poverty? The fact is that we automatically admit defeat in every quarter – whether it be the ravages of Nature, the extortions of our rulers, or the oppression of our *shastras* [scriptures]: whatever it is, we lack the strength to resist it.[21]

Others, too, reflected on the sorry state of Bengal, whether to blame the Bengalis' own inertia and defeatism or to contrast Bengal's present weak and sickly condition with a pre-colonial

---

[20]   *Glimpses of Bengal: Selected Letters of Rabindranath Tagore*, trans. Krishna Dutta and Andrew Robinson, London, 1991, p. 111.
[21]   Ibid., p. 110.

golden age. Speaking at a sanitary conference in 1912, Motilal Ghosh, a Calcutta newspaper editor, claimed that sixty years earlier the Bengali countryside had been remarkably healthy and disease-free: fever was rare, cholera practically unknown and smallpox subdued. In those days, the 'pick of the nation' lived in the countryside. Villages teemed with 'healthy, happy and robust people', who were untroubled by 'the bread question or the fear of being visited by any deadly pestilence'. But, Ghosh claimed, those idyllic days were gone. He dated the 'deterioration of the race' to an outbreak of 'Burdwan fever' (malaria) in the 1860s. 'Within the last sixty years,' he said, citing official statistics as evidence, 'malaria and cholera have swept away tens of millions of people from Bengal. Those who have been left behind . . . are more dead than alive.' Where once health and contentment had reigned, now there was hunger and villagers were 'dying like flies' from malaria. Ghosh concluded that the Bengali race was dying out, and 'must ultimately disappear like the old Greeks . . . unless vigorous steps of the right sort are promptly taken to save them from extinction.'[22]

Despite the material (as well as moral) progress said by the British to have been made under their rule, to many Indians it appeared that the country and people had physically deteriorated under colonial rule. This was the *Kaliyuga*, the age of affliction. It was as if India, visited as it was in the 1890s and 1900s by epidemics of malaria and plague and by one of the worst famines on record, were in the same deplorable position in which Europe had been five hundred years earlier at the time of the Black Death. But at least then Europe had been free to work out its own environmental destiny. India, a colony, could not. Only under *swaraj*, self-rule, it was widely believed, would India's blighted landscape smile once more.

[22] *Proceedings of the Second All-India Sanitary Conference*, vol. II, Simla, 1913, pp. 514–23. For the medical evidence behind Ghosh's pessimism, see Klein, 'Malaria and mortality in Bengal'.

Conclusion

'The sacred word "nature",' it has been said, 'is probably the most equivocal in the vocabulary of the European peoples.'[1] Certainly, historians have not found nature easy to handle either as a concept or as a historical influence. Many have responded by quietly ignoring it, believing the true subject of history to lie elsewhere – in the study of humankind alone, by nature unadorned and unaffected. And yet nature, or the environment as we more commonly describe it today, cannot so easily be brushed aside. As we have seen in this book, where historians might fear to tread, others – philosophers, geographers, naturalists, polemicists of all descriptions – have been ever ready to rush in. Whether historians like it or not, ideas of nature have played a major, indeed one might say integral, part both in the processes of history and in its interpretation. Moreover, these days, in an age of increasing environmental awareness, historians cannot, and should not, be silent on a subject of such broad interest and legitimate public concern.

But there has clearly been no consensus among those historians who have engaged with nature as to what it should signify. For some, nature represents an often harrowing array of environmental factors – climate, disease, dark and gloomy forests – which, to

[1] Arthur O. Lovejoy and George Boas, *Primitivism and Related Ideas in Antiquity*, Baltimore, 1935, p. 12.

varying degrees, are seen to have helped shape the course of human history. For others, nature is less material than perceptual, a way past peoples understood the world or privileged one kind of landscape over another. In actuality, as the later chapters of this book have tried to show, it is difficult to separate the two. Nature and culture are so closely entangled, it would often be foolish (and historically misleading) to try to separate them. But it is striking how commonly, in the discussion of material nature, writers about history (not always historians) espouse, or perhaps unconsciously adopt, some form of geographical or biological determinism. And it is striking, too, how often two of the most influential figures in environmentalist thought – Malthus and Darwin – inform (and continue to inform) such interpretations. Nature, as embodied in science, becomes a source of authority and one from which often sweeping historical judgements are made. History has learnt much from science over the past two centuries, almost as much as science has learnt from history. But it is surely salutary to see science itself as a cultural construction and to remember the extent to which our views, like those articulated by people in the past, are inescapably the product of the world-view of a particular age, society, class or gender.

The environmentalist paradigm – the idea that the environment has been a powerful force in human history and, conversely, that humankind has played a major part in refashioning nature – has been employed in different ways and at different times. At times, climate has been the dominant idiom; at others, health and disease, or the 'imperialism' of plants and animals. For some historians, environmentalism uniquely opens up an understanding of history as a global phenomenon as if viewed from space: a history of the world's unification by disease, for its remorseless transformation through deforestation and settled agriculture. But environmentalist ideas have also been invoked to explain difference and to divide up the world judgementally between the chosen and the damned – between nations and races, between such broad aggregations of people and places as the Orient and the West, the 'Neo-Europes' and tropics, or simply between civilization and savagery.

Williams, *The Country and the City* (St Albans, Paladin, 1973), and Carolyn Merchant, *The Death of Nature: Women, Ecology and the Scientific Revolution* (San Francisco, Harper, 1983). Also, from a gender perspective, see the essays in Carol P. MacCormack and Marilyn Strathern (eds), *Nature, Culture and Gender* (Cambridge, Cambridge University Press, 1980), and for the disease connection, James C. Riley, *The Eighteenth-century Campaign to Avoid Disease* (Basingstoke, Macmillan, 1987).

An important new historical perspective is provided by recent interpretive studies of landscape, including Denis E. Cosgrave, *Social Formation and Symbolic Landscape* (London, Croom Helm, 1984); Michael Conzen (ed.), *The Making of the American Landscape* (Boston, Unwin Hyman, 1990); D. W. Meinig (ed.), *The Interpretation of Ordinary Landscapes: Geographical Essays* (New York, Oxford University Press, 1979); John Rennie Short, *Imagined Country: Environment, Culture and Society* (London, Routledge, 1991); and, from a more specifically historical perspective, Simon Schama, *Landscape and Memory* (London, HarperCollins, 1995). A more materialistic interpretation of the relations between environment, society and history is to be found in such works as Robert I. Rotberg and Theodore K. Rabb (eds), *Climate and History: Studies in Interdisciplinary History* (Princeton, NJ, Princeton University Press, 1981), and I. G. Simmons, *Changing the Face of the Earth: Environment, History, Culture* (Oxford, Blackwell, 1989). Anthropological studies of landscape meaning and human ecology have not been discussed in this book, but for an introduction to some of the issues, see Elisabeth Croll and David Parkin (eds), *Bush Base, Forest Farm: Culture, Environment and Development* (London, Routledge, 1992), especially the article 'Culture and the perception of the environment' by Tim Ingold. For a more narrowly focused approach, see Robert Harms, *Games against Nature: an Eco-cultural History of the Nunu of Equatorial Africa* (Cambridge, Cambridge University Press, 1987).

On the historiographical front, the work of the *Annales* School is succinctly discussed in Peter Burke, *The French Historical Revolution: the Annales School, 1929–89* (Cambridge, Polity Press, 1990). There are biographies of several of the historians and geographers discussed in chapters 2 and 3 of this book. Among the most useful are Carole Fink, *Marc Bloch: a Life in History* (Cambridge, Cambridge University Press, 1989), and W. H. McNeill, *Arnold J. Toynbee: a Life* (New York, Oxford University Press, 1989). Geoffrey J. Martin's *Ellsworth Huntington: his Life and Thought* (Hamden, Conn., Archon Books, 1973) is disappointing: a far more incisive account of this influential geographer is to be found in David N. Livingstone, 'Climate's moral economy: science, race and place in post-Darwinian British and American geography', in Anne Godlewska and Neil Smith (eds), *Geography and Empire* (Oxford, Blackwell, 1994), pp. 132–54.

A steadily growing number of works deal with environmentalist ideas in the context of particular countries and periods. Among the most insightful and accessible of these are D. G. Charlton, *New Images of the Natural in France: a Study in European Cultural History, 1750–1800* (Cambridge, Cambridge University Press, 1984); Keith Thomas, *Man and the Natural World: Changing Attitudes in England, 1500–1800* (Harmondsworth, Penguin, 1984); and Roderick Nash, *Wilderness and the American Mind* (3rd edn, New Haven, Conn., Yale University Press, 1967). The American literature is especially rich and innovative. Among a host of studies, two are particularly enlightening: William Cronon, *Changes in the Land: Indians, Colonists and the Ecology of New England* (New York, Hill and Wang, 1983), and Donald Worster, *Rivers of Empire: Water, Aridity and the Growth of the American West* (New York, Oxford University Press, 1985).

The environmental history of regions beyond Europe and North America is also developing rapidly, and the work of Crosby, Curtin and Grove has been a major stimulus to further research and debate in several fields. Some of the most important work has so far appeared in collections of essays, such as the volume, already cited, on *Geography and Empire*, edited by Anne Godlewska and Neil Smith (Oxford, Blackwell, 1994), and John M. MacKenzie (ed.), *Imperialism and the Natural World* (Manchester, Manchester University Press, 1990). In addition to works cited in chapters 8 and 9 of this book, significant studies of environment and history in Africa include David Anderson and Richard Grove (eds), *Conservation in Africa: People, Policies and Practice* (Cambridge, Cambridge University Press, 1987); the articles on 'The politics of conservation in southern Africa' in a special issue of the *Journal of Southern African Studies*, 15 (2), 1989; John M. MacKenzie, *The Empire of Nature: Hunting, Conservation and British Imperialism* (Manchester, Manchester University Press, 1988); and Michael A. Osborne, *Nature, the Exotic, and the Science of French Colonialism* (Bloomington, Ind., Indiana University Press, 1994). Helge Kjekshus, *Ecology Control and Economic Development in East African History: the Case of Tanganyika, 1850–1950* (London, Heinemann, 1977), and John Iliffe's *A Modern History of Tanganyika* (Cambridge, Cambridge University Press, 1979), are two interesting earlier examples of the effective use of environmental perspectives.

For Asia, too, a large number of regional studies have now been published, both about indigenous ideas and practices and those of European colonial powers. On the former, the essays edited by Ole Brunn and Arne Kalland, *Asian Perceptions of Nature* (Copenhagen, Nordic Institute of Asian Studies, 1992) are a valuable corrective to simplistic views of Asian attitudes to the environment, while Victor R. Savage, *Western Impressions of Nature and Landscape in Southeast Asia* (Singapore, Singapore University Press, 1984), perhaps best read alongside S. H. Alatas's *Myth of the Lazy Native* (London, Cass, 1977), provides for one region of Asia a useful account of European perspectives. For India, in addition to Madhav Gadgil and Ramachandra Guha's *This Fissured Land* (Delhi, Oxford University Press, 1992), recent studies include David Arnold and Ramchandra Guha (eds), *Nature, Culture, Imperialism: Essays on the Environmental History of South Asia* (Delhi, Oxford University Press, 1995), and R. H. Grove and V. Damodaran (eds), *Essays on the Environmental History of South and Southeast Asia* (Delhi, Oxford University Press, 1996).

Of necessity, this book has had to be selective with respect to themes covered and regions discussed. One of the most glaring omissions must be the environmental history of Australia, about which there are now a number of fascinating studies. These include Paul Carter, *The Road to Botany Bay: an Essay in Spatial History* (London, Faber and Faber, 1987); J. Powell, *Environmental Management in Australia, 1788–1914* (Melbourne, Oxford University Press, 1976); Stephen Dovers (ed.), *Australian Environmental History: Essays and Cases* (Melbourne, Oxford University Press, 1994); and, for an indigenous understanding of the landscape, Ronald M. Berndt and Catherine H. Berndt, *The Speaking Land: Myth and Story in Aboriginal Australia* (Ringwood, Victoria, Penguin, 1989). Finally, mention must be made of one of the most original and influential of all works linking art, environment and the history of ideas, Bernard Smith's *European Vision and the South Pacific* (2nd edn, New Haven, Conn., Yale University Press, 1985).

Index

Aborigines, 28, 88, 90, 121
Adams, Herbert Baxter, 100
Amazon, 35, 142, 156
Americas, Spanish invasion and
  conquest of, 78–85, 89, 126,
  130–1, 176
Amerindians, 77–85, 89, 90, 92, 94,
  96, 101–3, 112–15, 117, 121–2,
  124, 125–7, 128, 129, 137; *see
  also* Arawaks; Aztecs; Incas
Anderson, Benedict, 185
Annales School, 39–47, 57, 120
anti-Semitism, 34, 66, 67
Arawaks, 78, 125–6, 144
Aristotle, 17
Ashburn, P. M., 90
Assam, 170, 172, 182
Athens, 62
Australia, 29, 87, 88, 89, 107, 110,
  122
Aztecs, 78, 80, 81, 82, 85, 131

Bahamas, 33
Banks, Sir Joseph, 164, 165
Barbados, 93, 94, 128, 153
Bengal, 153, 154, 169, 185–7
Beowulf, 134
Bernier, François, 23

Bible, 58, 85, 134
Black Death, 8, 45, 58, 62–7, 75,
  78, 79, 80, 83, 85, 92, 99, 110,
  187; *see also* plague
Bligh, William, 166
Bloch, Marc, 39, 40, 41, 42, 43, 44,
  45, 64
Bodin, Jean, 23
Bondt, Jacob, 164
Borah, W. W., 78, 83, 125
Bougainville, Louis Antoine de, 48,
  150
Bradford, William, 134
Braudel, Fernand, 39, 40, 41–4, 45,
  46, 47, 57, 111
Brazil, 28, 87, 93, 111, 127, 148,
  155, 156, 161–2, 169
Bristol, 63
Britain, 44, 64, 93, 171, 172
Brockway, Lucile H., 166–7
Brooks, Francis J., 84–5
Bryant, William Cullen, 139
Bryson, Alexander, 155
Buckle, H. T., 25, 30
Buddhism, 132
Buffon, Georges, 138, 164
Burke, Edmund, 155
Burma, 69, 169, 181

Calcutta, 153, 181, 187
California, 102, 115
Cambodia, 35, 37
Canada, 36, 107, 116, 122, 123
Canary Islands, 80, 86–7, 127
Cape of Good Hope, 89, 165
Caribbean, see West Indies
Caribs, 125
Carson, Rachel, 2, 131
Casas, Bartolomé de las, 79, 127
cattle, 80, 87, 127, 129, 131
Ceylon, 37, 145, 156–7, 169
Chamberlain, Houston Stewart, 34
Chardin, Sir John, 23
China, 21, 22, 24, 36, 69, 70, 75,
    136, 165, 166
cholera, 75, 171, 172
Christianity, 130, 131, 132, 135
Church, Frederic, 147
cinchona, 77, 165, 166, 167, 171
Cole, Thomas, 139, 146
Columbus, Christopher, 6, 8, 44,
    48, 61, 76, 80, 84, 89, 109, 112,
    127, 136, 143–4, 146
Commerson, Philibert, 48
Congo, 35, 142, 150
Conrad, Joseph, 150
Constable, John, 137
Constantinople, 62, 66
Cook, James, 20, 147, 150
Cook, S. F., 78, 83, 125
Cooper, James Fenimore, 139
Cortes, Hernan, 78, 82, 87, 88,
    126, 130
Cronin, William, 121–2
Crosby, Alfred W., 5, 76–7, 80, 84,
    86–91, 96, 98, 108, 109, 122,
    129, 130, 131, 176
Curtin, Philip D., 145, 158

Darwin, Charles, 26–7, 30, 50–1,
    56, 89, 132, 148–9, 155, 161,
    189

Darwinism, 30, 31, 37, 58, 90, 96,
    99, 103, 105, 113, 114, 122, 174
Defoe, Daniel, 150
Disraeli, Benjamin, 27

East India Company (English), 165,
    177
East India Company (Swedish), 165
Easter Island, 54–5
'ecological imperialism', 5–6, 86–7,
    88, 99,
Eden, 48, 50, 134, 135, 142, 144,
    148, 149, 150, 164, 169
Egypt, 17, 22, 24, 37, 62
Encyclopédie, 24ᵃ
Engels, Friedrich, 24
England, 51, 65, 71, 122, 128, 137

famine, 71, 172, 177, 178, 187
Febvre, Lucien, 39, 41
Fernando Po, 155
Florence, 63, 65
forests, 40, 41, 51–2, 113, 121,
    124–5, 133, 134, 179–83
France, 39, 40–1, 43, 44, 45, 46,
    47, 57, 63, 64, 66, 71
Francis of Assisi, St, 131
Freyre, Gilberto, 161–2, 163
fur trade, 123–4

Gadgil, Madhav, 179, 183
Gait, E. A., 173
Gandhi, M. K., 185
Genoa, 63
Ghosh, Motilal, 187
Gilpin, William, 155
Glacken, Clarence J., 14, 17,
    19
Gobineau, Comte de, 34
Gottfried, Robert S., 62
Goubert, Pierre, 46
Graham, Maria, 161
Greece, 14, 17, 50, 169
Greeks, 15, 17, 18, 134, 187

Greenland, 61, 66
Grove, Richard H., 149, 152, 167, 183
Guha, Ramachandra, 179, 183
Guyana, 150, 171

Haeckel, Ernst, 3
Haiti, 78, 95
Hegel, G. W. F., 24, 136
Herodotus, 17, 35, 136
Hillary, William, 151
Hinduism, 132, 171, 185
Hindus, 174
Hippocrates, 14–18, 19, 35, 37, 54, 89, 136, 173
Hippocratic ideas, 20, 21, 54, 152
Hispaniola, 78, 80, 82, 84, 95, 127
Hofstadter, Richard, 106
Holdich, T. H., 172
horses, 77, 82, 87
Hoskins, W. G., 136
Hudson River School, 138–9, 147
Hudson's Bay Company, 123
Humboldt, Alexander von, 50, 56, 137, 146–9, 157, 160, 169
Hunter, John, 151
Huntington, Ellsworth, 31–4, 35, 43, 46, 53, 60, 105, 155, 159, 171
Huxley, Aldous, 157

Iceland, 61, 62
Incas, 36, 80, 82, 131
India, 21, 23, 24, 35, 70, 75, 153, 160, 166, 167, 168, 169–87
influenza, 75, 77, 92, 179
Ireland, 172
irrigation, 177–8
Islam, 21, 67, 134
Italy, 63, 137

Jackson, Andrew, 104
Jamaica, 151
Jefferson, Thomas, 138

Jesuits, 124, 165
Jews, 66, 79
Jones, Sir William, 169
Joralemon, Donald, 83
Judaeo-Christian tradition, 6, 131–2, 133

Kansas, 117, 118
Kerala, 169, 170
Kew Gardens, 164, 165, 166, 167
Kiple, K. F., 95
Knox, Robert, 27

Labrousse, Ernest, 46
Le Roy Ladurie, Emmanuel, 39, 45–7, 71, 72, 74, 75, 80
Lind, James, 151, 152, 153, 154
Linnaeus, C., 164, 165
Lisbon, 59
London, 63, 64
longue durée, 39, 40, 43, 44–5, 47, 58, 88, 120
Lovelock, James, 56–7
Lyell, Charles, 26

McNeill, W. H., 69, 74, 76, 77, 79, 80, 82, 90, 129
Madeira, 86
malaria, 32, 93, 94, 95, 96, 163, 165, 174–5, 178, 179, 180, 187
Malin, James C., 114, 116–18, 119
Malthus, T. R., 49–51, 53, 54, 189
Malthusianism, 47, 54, 55, 56, 71–2, 73, 80, 83, 92, 110, 111, 118, 149
Maoris, 28, 88, 90
Marchy, R. B., 117
Marseilles, 64, 73
Marsh, George Perkins, 52, 125, 149
Marshall, Henry, 156
Martin, James Ranald, 153
Marx, Karl, 13, 24
Mauritius, 149

Mayan civilization, 32, 35, 36, 37, 160
measles, 62, 74, 75, 77, 81, 92
Mexico, 35, 78, 80, 82, 83, 84, 87, 115, 120, 126, 127, 130, 146
Mongols, 69–70, 75
Montaigne, M., 137
Montesquieu, Baron de, 20–4, 26, 30, 31, 32, 33, 35, 46, 50, 89, 98, 136
Montezuma, 82
Moseley, Benjamin, 152
Mughals, 171, 180

Napoleon Bonaparte, 95
Nashy, Roderick, 133–4, 135
native Americans, see Amerindians
'Neo-Europes', 7, 87, 88, 94, 108, 109, 162, 176, 189
New England, 121–2, 127, 134, 147
New Zealand, 28, 75, 87, 88, 110, 122

d'Orta, Garcia, 164

Palmer, Samuel, 137
Paris, 163, 164
Parkman, Francis, 102
Peru, 80, 82, 130, 167
pigs, 77, 80, 86, 87, 126–7, 129
Pilgrim Fathers, 133, 139
Pizarro, Francisco, 82
plague, 62–73, 74, 75, 80, 83, 92, 145, 163, 179, 187; see also Black Death
Poland, 63, 66
Ponting, Clive, 53–5
Portugal, 92
Post, John D., 59–60, 61
Pratt, Mary Louise, 146
Punjab, 177, 178

quinine, 93, 167; see also cinchona

race, 13, 26–9, 33–5, 90, 152, 154–5, 159
railways, 178–9, 181
Raleigh, Sir Walter, 150
Rohrbough, Malcolm J., 120
Romanticism, 19, 137, 139, 146, 148, 186
Rome, 32, 50, 62, 139, 169
Ross, Ronald, 174–5, 176
Rousseau, Jean-Jacques, 19, 101, 137
Russia, 66

Said, Edward W., 141
St Helena, 149
Sauer, Carl O., 84, 114–16, 119, 120, 135
Semple, Ellen Churchill, 30–1, 33, 43, 105, 154, 171
sheep, 80, 87, 127, 130, 131
Siberia, 77, 123
Sienna, 63
slavery, 26, 92–7, 101, 128, 143, 160–1
slaves, 66–7, 91, 94, 95, 126, 128, 152, 166, 171
Sloane, Hans, 151
smallpox, 62, 74, 75, 77, 80, 81, 82, 84, 87, 88, 92, 96, 129, 171
Smeathman, Henry, 145, 158
Smith, Adam, 24
South Africa, 107, 110, 122
Spain, 30, 66, 130–1, 140
Sri Lanka, see Ceylon
Stannard, David E., 85
Stannus, H. S., 153
Stebbing, E. P., 182
Steuart, James, 158
sugar, 87, 92–3, 94, 96, 126–8, 153, 165
Swettenham, Frank, 159
syphilis, 77, 163, 164, 176

Tagore, Rabindranath, 186
Tahiti, 48, 50, 54, 145, 150, 166

Tasmanians, 91
Tavernier, François, 23
Tenochtitlán, 82, 85, 87
Thoreau, Henry David, 139
Thorne, Robert, 144
Toynbee, Arnold, J., 34–8, 47, 53,
    54, 67, 69
Trapham, Thomas, 151
Tropicalista School, 161
Turner, Frederick, 133
Turner, Frederick Jackson, 99–108,
    109, 111–16, 118–21, 130, 132,
    133, 137
Turner, J. M. W., 137
typhus, 77, 92

United States, 51, 93, 100–2, 105,
    107, 109, 114, 125, 138, 185
Uppsala, 163, 164, 165
Utterström, Gustaf, 46

Vasco da Gama, 176
Venice, 73
Vidal de la Blache, Paul, 39, 40
Virginia, 127, 138
Voltaire, François, 59

Wallace, Alfred Russel, 28–9, 89, 156
Watts, David, 125, 126
Webb, Walter Prescott, 109–12, 113
West Indies, 48, 78, 80, 84, 92–4,
    95, 96, 122, 125–8, 129, 143–4,
    147, 151–2, 155, 160, 166, 169
White, Gilbert, 57, 59
White, Lynn, 131–2
Wigglesworth, Michael, 134
Williams, Michael, 113, 124
Worster, Donald, 20, 50

yellow fever, 77, 93, 94–5, 96, 128

Ziegler, Philip, 72